DATA BANK ON GENERAL SCIENCE

DATA BANK
ON
GENERAL SCIENCE

Compiled and Edited by
Arvind M. Dwivedi

ANMOL PUBLICATIONS PVT. LTD.
NEW DELHI - 110 002 (INDIA)

ANMOL PUBLICATIONS PVT. LTD.
4374/4B, Ansari Road, Daryaganj
New Delhi-110 002
www.anmolbooks.com

Data Bank on General Science

First Edition 2001
ISBN 81-261-1000-7

PRINTED IN INDIA

Published by J. L. Kumar for Anmol Publications Pvt. Ltd., New Delhi-110 002 and Printed at Mehra Offset Press, Delhi.

CONTENTS

PREFACE

The present age is the age of competitions. Gone are the days when the degrees were enough to get a job. Now the thorough study of any subject is required to get through in any competitive examination. The objective multiple choice type questions are asked almost in all the competitions. In order to meet the increasing demand of the aspirants, we have prepared the Data Bank on General Science and feel pleasure on presenting the same before you. It is concise, comprehensive and dependable which has a wide coverage of the syllabus.

The proportion of questions from Science in General study has get tremendous increase in the recent past. Near about 15% questions are asked from this portion. In the present data bank we have touched all the corners of syllabus as well as the latest pattern of questions asked in UPSC, SSC, LIC, GIC, CDS, NDA and other examinations. We have taken utmost care in preparing the questions with particular gradation and level of the various examination.

We shall feel grateful to the readers if they send us their valuable comments so that further improvements may be made in the subsequent editions of the book.

—Arvind M. Dwivedi

1
GENERAL PHYSICS

1. Image formed by a plain mirror is always
   (a) Virtual (b) Real
   (c) Bigger in size (d) Smaller in size.

2. A body is moving with a uniform speed. The acceleration of the body is
   (a) Zero (b) Positive
   (c) Negative (d) Constant.

3. The supporting or balancing point about which a lever moves is called the
   (a) Centre
   (b) Apex
   (c) Point of application of land
   (d) Fulcrum.

4. Rainbow is normally formed due to
   (a) Reflection of light
   (b) Refraction of light
   (c) Total internal refraction of light
   (d) Diffraction of light.

5. In max-min thermometer mercury is one of the liquids the second liquid is
   (a) Spirit (b) Alcohol
   (c) Water (d) Kerosene.

6. We receive the sun rays by the process of object is called
   (a) Conduction (b) Convection
   (c) Radiation (d) None of these.

7. The instrument used to see a tiny object is called
   (a) Oscilloscope (b) Telescope
   (c) Periscope (d) Microscope.

8. Rate of change of momentum is called
   (a) Acceleration (b) Force
   (c) Power (d) Energy.

9. One horse power is equal to how many watts
   (a) 526 (b) 726
   (c) 780 (d) 746.

10. Power of the lense is measured in
    (a) Horse power (b) Light year
    (c) Diopter (d) Watt.

11. How much work should be done to produce one calorie of heat?
    (a) 4.2 joules (b) 4.2×10^2 joules
    (c) 42 joules (d) None of these.

12. Unit of heat is
    (a) Ampere (b) Centigrade
    (c) Joules (d) Calorie.

13. Cathode rays consist of negatively charged particles which are known as
    (a) Electrons (b) Protons
    (c) Neutrons (d) Deuterons.

14. The unit of current is
    (a) Ohm (b) Watt
    (c) Ampere (d) None of these.

15. Humidity is measured by
    (a) Hydrometer (b) Hygrometer
    (c) Barometer (d) None of these.

16. The Freezing point of water is
    (a) 4°c (b) 0°c
    (c) 100°c (d) None of these.

17. Which of the following can be made into a permanent magnet?
    (a) Soft iron (b) Hard steel
    (c) Lead (d) None of these.

18. The ratio voltage to current across a wire refers to
    (a) Conductance (b) Resistance
    (c) Electric power (d) None of these.

19. When a body is raised from the ground level stares
    (a) Potential energy (b) Kinetic energy
    (c) Heat energy (d) None of these.

20. The electrical energy is converted into mechanical energy by
    (a) Dynamo (b) Cell
    (c) Motor (d) None of these.

21. Lines of Magnetic force travel from north pole to
    (a) South pole (b) North pole
    (c) Both (a) and (b) (d) None of these.

22. Single pole magnet is
    (a) Not possible (b) Possible
    (c) Both (a) and (b) (d) None of these.

23. There is no atmosphere on moon there the sound will be
    (a) Listened (b) Not listened
    (c) Reached (d) Not reached.

24. The heater element is made of
    (a) Tungsten
    (b) Copper
    (c) Nichrome
    (d) None of these.

25. Sun rays in morning are less hot than at noon because
    (a) Sun is less hot in the morning
    (b) Earth absorbs the heat of the sun
    (c) Rays in the morning travel longer distance than at noon
    (d) None of these.

26. Jet aeroplanes fly in the principle of which law of motion?
    (a) First law
    (b) Second law
    (c) Third law
    (d) None of these.

27. Light Waves
    (a) Can travel in perfect vacuum
    (b) Require air to travel through
    (c) Require an electric field to travel through
    (d) None of these.

28. When a bird is flying high up it does not cast its shadow on the ground because
    (a) Light rays are not falling on it
    (b) The layers on atmosphere are dense
    (c) The size of sun is much bigger than bird
    (d) None of these.

29. In a barometer if water is used in place of mercury then the length of barometer tube has to be
    (a) Kept same
    (b) Increased
    (c) Decreased
    (d) None of these.

30. If two tubes of different diameters are dipped in water is
    (a) Greater in the tube of smaller radius
    (b) Greater in the tube of larger radius
    (c) Same in both tubes
    (d) None of these.

31. The density of water is maximum at
    (a) 0°c
    (b) 4°c
    (c) 2°c
    (d) 32°c.

32. The phenomenon of splilting of white light onto seven colours is known as
    (a) Diffraction
    (b) Interference
    (c) Dispersion
    (d) Total internal reflection.

33. Which of the following materials can be made into a permanent magnet?
    (a) Soft iron
    (b) Lead
    (c) Hard steel
    (d) Aluminium.

34. As we go up a hill the level of mercury in a mercury barometer
    (a) Increases
    (b) Decreases
    (c) Remains constant
    (d) Increases first and then decreases.

35. The sum of Kinetic and Potential energies of a falling body is
    (a) Constant at all points
    (b) Maximum in the beginning
    (c) Minimum in the beginning
    (d) Maximum in the middle of the path

36. The instrument used for detecting and measuring electric current is
    (a) Voltmeter
    (b) Galvanometer
    (c) Wattmeter
    (d) Anemometer

37. During the process of melting the heat supplied is utilised in
    (a) Decreasing the potential energy of the molecules
    (b) Increasing the Kinetic energy of the molecules
    (c) Increasing the potential energy of the molecules
    (d) None of these.

38. The capacity of the atmospheric air to retain water as vapour depends upon
    (a) Its pressure
    (b) Its density
    (c) Its temperature
    (d) The velocity of the wind.

39. A transformer is used to
    (a) Convert DC into a AC
    (b) Regulate the fluctuation of voltage
    (c) Measure electricity
    (d) Convert low voltage in to high voltage and vice versa

40. The mirror used in vehicle to see the rear view is
    (a) Convex mirror (b) Concave mirror
    (c) Plane mirror (d) None of these.

41. A piece of iron does not float on water because
    (a) It displaces no water
    (b) The mass of water displaced is less than that of iron
    (c) The mass of water displaced is equal to the mass of iron.
    (d) The mass of water displaced is more than that of iron.

42. The supporting or balancing point about which a lever moves is called
    (a) Centre (b) Fulcrum
    (c) Com (d) Apex.

43. A body which travels in space around a planet is
    (a) A Galaxy (b) A Phase
    (c) An Equator (d) Satellite.

44. A man carrying a bucket leans in order to
    (a) Equalise the weight of the bucket
    (b) Reduce the weight of the bucket
    (c) Reduce his muscle strain
    (d) Bring himself to stable eqilibrium.

45. Which of the following is an internal combustion engine
    (a) Steam engine (b) Gasoline engine
    (c) Dynamo (d) Storage battery.

46. For the same change in temperature the amount of expansion is maximum in
    (a) Gaseous state (b) Liquid state
    (c) Solid state (d) Dense state.

47. Sound travels faster in
    (a) Liquids (b) Solids
    (c) Gases (d) Vacuum.

48. Rainbow is normally formed due to
    (a) Reflection of light
    (b) Total internal reflection of light
    (c) Dispersion of light
    (d) Diffraction of light.

49. An electric current in telephone wires flow easily when
    (a) Carbon grains are not loosely packed
    (b) Carbon grains are not closely packed
    (c) The diaphragm is not fixed
    (d) You talk very loudly.

50. Water can be split into its components by
    (a) Electrolysis
    (b) Passing chlorine through it.
    (c) Adding acid
    (d) Heating.

51. An electric device which converts direct current into alternative current is known as
    (a) Aliminator (b) Generator
    (c) Rectifier (d) Inverter.

52. Barometer is used to measure
    (a) Temperature (b) Humidity
    (c) Atmospheric pressure (d) Rainfall.

53. The burns caused by steam are more severe than those caused by boiling water because
    (a) The temperature of steam is higher than that of boiling water
    (b) The steam is in vapour state
    (c) Steam everts more pressure
    (d) Of latent heat of steam.

54. Who invented the telescope?
(a) James Watt (b) Galilio
(c) J.L.Bairol (d) Rotengen.

55. Instrument used to measure radioactivity is
(a) Radio micrometer (b) Radio graph
(c) radio compass (d) Geiger counter.

56. Sun is constantly dealing energy due to
(a) Nuclear fission (b) Nuclear fusion
(c) Radiation (d) None of the above.

57. Which of the following is the highest particle
(a) Proton (b) Neutron
(c) Electron (d) Atom.

58. Which of the following is non-magnetic?
(a) Cobalt (b) Nickel
(c) Iron (d) Brass.

59. Like poles always
(a) Attract each other
(b) Repel each other
(c) Both (a) and (b)
(d) Some time (a) and some time (b).

60. 1 light year is equal to
(a) 10^{15} m (b) 9.46×10^{15} m
(c) 9.46×10^{15} km (d) 9.46×10^{15} cm.

61. Time taken by the sunlight to reach the earth is
(a) 10^2 minute (b) 8 minutes
(c) 10^4 sec (d) 10^5 sec.

62. 1 metre is equal to
(a) 10 decimetre (b) 10 decametre
(c) 3 feet (d) 100 millimetre.

63. 1 kilogram weight is equal to the
(a) 1000 gm (b) 1000 gm weight
(c) 100 mg (d) 9.8 Newtons.

64. One horse power is equal to
    (a) 746 watt (b) 736 watt
    (c) 756 watt (d) 746 watt.

65. Which of the following statements is correct ?
    (a) Solids cannot be deformed
    (b) Gases cannot be pressurised
    (c) Liquids do not expand to fill their containers
    (d) All gases are collected easily.

66. The energy possessed by the body lying on the top of the out at minar is
    (a) Kinetic energy (b) Automatic power energy
    (c) Gravitational energy (d) Potential energy.

67. The energy possesed by a man running on a level track is
    (a) Gravitational energy
    (b) Kinetic cum-potential energy
    (c) Track energy
    (d) Kinetic energy.

68. Why is it that a car turning forward on a cinema screen the wheels often appear to be turning backward?
    (a) The speed of change of frames in the film is very slightly greater than the speed of revolution of the car wheels
    (b) Due to the phenomena of velocity
    (c) Due to the angular velocity
    (d) Due to resistance of vision.

69. The pulling of the cart by the horse is due to the Force
    (a) That the cart exerts force on the horse
    (b) That the horse exerts a force on the cart
    (c) That the horse exerts a force on the ground
    (d) That the ground exerts a force on the horse.

70. A beam balance is asked for comparing the
    (a) Densities of two bodies
    (b) Weights of two bodies

(c) Masses of two men
(d) None of these.

71. A bomb is dropped from an aeroplane flying horizontally at constant speed where will the aeroplane be when the bomb hits the ground?
(a) Considerably behind it
(b) Over the place
(c) Near the place
(d) Considerably farther from the spot

72. Which of the rain drops falls faster big one or the little one?
(a) Big one
(b) Little one
(c) Both will fall equally fast
(d) None of the above.

73. If a ship passes from fresh water into sea water it will
(a) Remain unaffected (b) Not sink
(c) Rise higher (d) Sink lower.

74. If cream be removed from milk the density of left milk
(a) Will increase (b) Will decrease
(c) Will remain the same (d) Will be higher.

75. The temperature on which the value of Centigrade and Fahrenheit is equal
(a) 40° C (b) 40° C
(c) 40° F (d) - 40° C.

76. The melting and freezing point of ice is
(a) 0° C and -0° C
(b) -0° Centigrade and -0° Centigrade
(c) 0° F and 0° C
(d) only zero degree Centigrade.

77. Heat can be obtained from
(a) Sun (b) Electricity
(c) Friction (d) All the above.

78. Boiling point of water at sea level is
    (a) 10°C (b) 10°C
    (c) 100°F (d) 100°C.

79. Temperature can be measured by
    (a) Caloriemeter (b) Manometer
    (c) Thermometer (d) Radiometer.

80. Which thermometer is used by us to read the temperature of living bodies?
    (a) Alcohol (b) Mercury
    (c) Wine (d) Water.

81. Minimum or maximum temperature of the day can be measured
    (a) By Centigrade thermometer
    (b) Only the sicks thermometer
    (c) Only the gas thermometer
    (d) By clinical thermometer.

82. Evaporation of water caused by the heat of the
    (a) Sun (b) Volcanoes
    (c) Earth (d) Stars.

83. The normal body temperature of a human body is
    (a) 98.0 degree Farenhight
    (b) 95.4 degree Centigrade
    (c) 98.4 degree Centigrade
    (d) 36.9. degree centigrade.

84. Upper fixed point of the thermometer is the temperature at which
    (a) Water freezes (b) Water boils
    (c) Mercury boils (d) Ice melts.

85. The surface of earth is just going to freez what will be the temperature at the bottom
    (a) Ten degree centigrade
    (b) Four degree centigrade

(c) Minus four degree centigrade
(d) Minus two degree centigrade.

86. The Rocket moves up the consuming the
(a) Nuclear energy (b) Solar energy
(c) Gravitation energy (d) Sun energy.

87. The alternative source of heat on the surface of the earth is
(a) Nuclear energy (b) Gravitational energy
(c) Solar energy (d) Radiation energy.

88. The boiling point of a liquid when boiled at higher altitudes will
(a) Increase
(b) Decrease
(c) Decrease or increase according to the quality of liquid
(d) Remain same.

89. Heat is most closely related to
(a) Temperature (b) Energy
(c) Momentum (d) Friction.

90. What instrument is used to measure the humidity of air?
(a) Hygrometer (b) Hydrometer
(c) Barometer (d) Opisometer.

91. Unit of heat is
(a) Degree (b) Calorie
(c) Centigrade (d) Kelvin.

92. Can water boil without being heated?
(a) No , it may not be
(b) Yes, it may be by changing the pressure
(c) Yes, it may be
(d) Yes, it may be changed by changing temperature.

93. Heat and temperature are related as
(a) Rain and water simultaneously
(b) Power and shining
(c) Electricity and current
(d) Sugar and sweetness.

94. We can measure with the help of thermometer
   (a) Heat of the substance
   (b) Temperature of the substance
   (c) Constant temperature
   (d) Pressure of the substance.

95. One should not dip a doctors thermometer into hot water because
   (a) It will break into small pieces
   (b) It will become hot
   (c) It will be stout
   (d) Mercury will come out of it.

96. We generally use black cloth for umbrella, because
   (a) It absorbes heat
   (b) Heat comes out of it without wastage
   (c) Heat crosses out of it
   (d) Heat is reflected from it.

97. An ideal gas thermometer uses
   (a) Air (b) Alcohol
   (c) Hydrogen (d) Mercury.

98. Caloriemeter is made of copper because
   (a) Specific heat of copper becomes less
   (b) Specific heat of copper becomes high
   (c) Copper is good conductor of heat
   (d) Copper is available easily.

99. The freezing point of a substance is always lower than of its
   (a) Melting point (b) Boiling point
   (c) Cooling point (d) Specific heat.

100. At what temperature water occupies the minimum volume?
   (a) Zero degree Centigrade
   (b) Twenty degree centigrade
   (c) Two degree centigrade
   (d) None of these.

101. Transformation of heat involves

(a) Single mode (b) Double mode
(c) Three types of mode (d) Four modes.

102. Metals are heated

(a) By conduction
(b) By convention
(c) By radiation method
(d) By radiation and convection.

103. Which of the engines has more capacity?

(a) Vapour (b) Petrol
(c) Fuel (d) Diesel.

104. Which of the following engines is used in car and aeroplane?

(a) Heat engine (b) Diesel engine
(c) Petrol engine (d) Atomic engine.

105. Hailstones are formed when

(a) Water freezes (b) Rain drops freezed
(c) Air freezes (d) Cloud freezes.

106. The instruments used to study the properties heat radiation is

(a) Thermopiles
(b) Ritcher apparatus
(c) Differential air thermoscope
(d) Monometer.

107. Unit of time period is

(a) Second (b) Vibration / second
(c) Second / vibration (d) Hour.

108. Movement of the pendulam of a clock is called

(a) Motion
(b) Movement
(c) Simple harmonic motion
(d) Simple vibration.

109. Sound is a form of
(a) Electrical energy (b) Heat energy
(c) Mechanical energy (d) Potential energy.

110. Sound does not travel in
(a) Iron rod (b) Water
(c) Vacuum (d) Hydrogen gas.

111. The velocity of sound is greatest in
(a) Ammonia gas (b) Nitrogen gas
(c) Iron (d) Air.

112. Speed of light in comparison to sound is
(a) Greater (b) Lesser
(c) Equal (d) No difference.

113. The waves which vibrated in between 20 to 20,000 vib. per sound are called
(a) Sonic waves (b) Ultrasonic waves
(c) Infrasonic waves (d) Transverse waves.

114. Vibration more than 20,000 vib./sec. are known as
(a) Sonic waves (b) Ultrasonic waves
(c) Infrasonic waves (d) Transitory waves.

115. Limit audibility is
(a) 20 to 20,000 vibrations/seconds
(b) 20 vibrations/seconds
(c) 20,000 vibrations/second
(d) 40,000 or more vibrations/seconds.

116. Velocity of sound under normal condition is
(a) 332 metre per second
(b) 322.00 metre per second
(c) 342.00 metre per second
(d) None of the above.

117. Maximum velocity of sound is in the
(a) Summer season (b) Winter season
(c) Rainy season (d) Spring season.

118. Velocity of sound is the highest in
   (a) Solids (b) Liquids
   (c) Gases (d) Plasma.

119. Which of the following is true?
   (a) Earth is a transparent body
   (b) Sun is an opaque body
   (c) Moon is a shining body
   (d) Air is a transparent substance.

120. Light year in a unit of
   (a) Light (b) Sound
   (c) Heat (d) Distance.

121. Which of the following is non-luminous ?
   (a) Sun (b) Moon
   (c) Stars (d) Satellites.

122. Which of the following is an opaque body ?
   (a) Window glass (b) Brick
   (c) Glass plate (d) Water.

123. The sun light reaches the earth from sun is approximately?
   (a) 100 seconds (b) 1000 seconds
   (c) 1×10^{-5} seconds (d) 480 seconds.

124. The sun is a luminous body while the moon
   (a) Is a transparent body
   (b) Is a non-luminous body
   (c) Is a non-transparent body
   (d) Is an opaque body.

125. Which of the following is based on the principle:
   'Light travels in a straight line'
   (a) Projector (b) Pin hole camera
   (c) Magic lantern (d) Microscope.

126. The solar eclips which occurs after a years is called the
   (a) Solar lunar eclips (b) Annual eclips
   (c) Yearly eclips (d) None of these.

127. Which of the following mirrors should be used while shaving one's beard?

(a) Concave mirror (b) Convex mirror
(c) Plane mirror (d) Both (a) and (b).

128. The colour of the sky as seen from the outer space is

(a) Red (b) Blue
(c) White (d) Black.

129. Formation of shadows proves that

(a) Light travels in waves
(b) Light travels in a straight line
(c) Light travels (sometimes) in a parallel lines.
(d) Light travels in curved lines.

130. Which of the following is a false statement?

(a) Light is a form of enrgy
(b) Light can travel in a vacuum
(c) The velocity of light is infinitive
(d) The velocity of light in vacuum is 3×10^{10} cms per second.

131. The rays coming from the sun are

(a) Parallel (b) Transverse
(c) Longitudinal (d) None of these.

132. If a light ray falls on a plane mirror normally the angle of reflection will be

(a) 90° (b) 0°
(c) 45° (d) 180°

133. The mirror used in a motor car is

(a) Concave (b) Convex
(c) Plane (d) Both (a) and (b).

134. A diamond appears bright because

(a) It is radio active
(b) Of total internal reflection
(c) It has many facets
(d) Of high density.

135. Two plane mirrors are parallel to each other. A pen is placed in between them. The number of the images formed will be
(a) 6 (b) 1
(c) Infinite (d) Zero.

136. Which of the following is transparent?
(a) Candle (b) Sun
(c) Moon (d) None of the above.

137. Which of the following is transcluent?
(a) Greased paper (b) Wax
(c) Bricks (d) Milk.

138. Increase in the length of the day causes?
(a) Reflection of light by atmosphere
(b) Diffusion of light by atmosphere
(c) Total reflection of light
(d) Refrection of light from the upper layers of the atmosphere.

139. Rising sun's colour is red because
(a) Sun is red hot
(b) Sun is of red colour
(c) Sun has a reddish glow
(d) None of the above.

140. The sky appears blue because
(a) Of reflection of light (b) Of refraction of light
(c) Of polarisation of light (d) Of scattering of light.

141. Which one of the following is the least dispersed colour in the sun spectrum ?
(a) Voilet (b) Red
(c) Green (d) None of these.

142. In a film projector a lense is usually placed between the bulb and the film. This lense is a
(a) Convex lense
(b) Concave lens

(c) Plane lens
(d) Concave and convex lens.

143. Focal length of a concave lens is
(a) Positive (b) Negative
(c) +- both (d) No sign.

144. Focal length of a convex lens with the sign
(a) +- (b) -+
(c) + (d) -.

145. A fish is swimming in a long pond which has clear and calm water on looking towards the surface she might see
(a) Only the earth
(b) Only the sky
(c) Both the earth and the sky
(d) The sky and a mirror like reflection of the bottom of the whole pond.

146. Magnetic lines of force are
(a) Parallel to each other
(b) Perpendicular to each other
(c) Equivalant a like in all respects
(d) Paraller but opposite in direction.

147. Like poles
(a) Attract each other (b) Repel each other
(c) Remain uneffected (d) None of the above.

148. The opposite poles
(a) Attract each other (b) Repel each other
(c) Remain uneffected (d) None of the above.

149. Which of the following is a Ferrous metal?
(a) Zinc (b) Gold
(c) Silver (d) Ferrous.

150. The strongest attraction in a magnet is
(a) Near the ends (b) In the middle
(c) At the south pole (d) At the poles.

151. Which of the following is a non-magnetic metal?
(a) Cobalt (b) Copper
(c) Brass (d) Zinc.

152. To measure the magnetic field we use
(a) Voltmeter (b) Purometer
(c) Ventrimeter (d) None of the above.

153. Magnetic effect of current electricity was discovered by
(a) Einstien (b) Ampere
(c) Oerested (d) None of these.

154. Which of the following is non-magnetic?
(a) Nickel (b) Brass
(c) Cobalt (d) Iron.

155. An electromagnet is made from
(a) Soft iron (b) Steel iron
(c) Cobalt (d) Brass iron.

156. A strong magnet
(a) Repels iron and iron alloys
(b) Attracts some substances
(c) Attracts only iron and iron alloys
(d) Attracts all substances.

157. The number of magnetic elements of any place on the earth is
(a) 50 (b) 30
(c) 20 (d) None of these.

158. Unit of electric current is
(a) Volt (b) Joule
(c) Ampere (d) None of these.

159. In terms of electricity a solid conductor is one
(a) Which has no electrons
(b) Whose electrons are firmly bound
(c) Which has same electrons
(d) Some of whose electrons are firmly bound.

160. Electric current is

(a) Quantity of protons (b) Lack of electrons
(c) Quantity of neutrons (d) Excess of electrons.

161. Is it possible to measure the voltage of an A.C circuit with an electroscope?

(a) Yes (b) No
(c) It might may be (d) It might not be.

162. A body is projected upward with an initial velocity of 100 m/s. Find time when it will strike the ground.

(a) 10 sec (b) 20 sec
(c) 30 sec (d) 40 sec.

163. A wire of length L is stretched to twice its length that will be the change in resistance of the wire?

(a) 2 times (b) 4 times
(c) 6 times (d) 3 times.

164. The potential at all points inside a charged hallow sphere will be

(a) Zero
(b) Constant
(c) Proportional to the distance from centre.
(d) None.

165. The minimum energy required to extract an electron from an atom of a matal is termed as

(a) Work function (b) Threshold frequency
(c) Thresold wavelength (d) None.

166. The time period of pendulam is doubled when

(a) The length is doubled
(b) The mass of the bob is doubled
(c) The length becomes four times
(d) The mass of the bob becomes four times.

167. The curvature of a convex mirror is 20 cm. What is the position of the image of an object placed at a distance of 12cm from the mirror?

(a) 60/11 cm (b) 30/11 cm
(c) 11/60 cm (d) None.

168. Light travelling from vacuum enters water which of the following characteristic of light will remain in uncharged?
(a) Velocity (b) Amplitude
(c) Frequency (d) Wavelength.

169. Spherical aberration can be removed by using
(a) Convex lens (b) Concave lens
(c) Plano-convex lens (d) Cylindrical lens.

170. When white light is incident on a diffraction grating? The light that will be deviated from central image will be
(a) Yellow (b) Violet
(c) Indigo (d) Red.

171. The mass liberated during electrolysis depends on:
(a) Current
(b) Time for which current is passed
(c) Current as well as on time far which it is passed
(d) None.

172. A magnetic handle us kept in a non-uniform magnetic field it experiences
(a) A force and torque
(b) A torque but not a force
(c) Neither a force nor a torque
(d) A force but not a torque.

173. On increasing the number of electrons striking the anode of an X-rays tube, which of the following characteristic of the X-rays increase?
(a) Frequency (b) Intensity
(c) Wavelength (d) Power.

174. The half life of randon is 3.8 days. After how many days will only one twentieth of redon sample be left over.
(a) 17.45 days (b) 16.45 days
(c) 15.45 days (d) None.

175. Electrons can be accelerated to very high energies by means of
   (a) Cyclotrons (b) Behatrons
   (c) Thyratrons (d) None of these.

176. The magnitude of fields vector H at a radius at 1 metre from a long conductor is 2 amp/metre. What will be the current in the wire?
   (a) 4 II amp (b) 3 II amp
   (c) 2I amp (d) None of these.

177. The induced e.m.f. in a conductor is
   (a) Inversely proportional to the rate of change of flux
   (b) Directly proportional to the rate of change of flux
   (c) Directly proportional to the total of conductor
   (d) None of these.

178. At what height above the earth's surface has diminised by one tenth of one percect
   (a) 3.2 k.m (b) 2.3 k.m
   (c) 3 km (d) None.

179. The density of atmoshphere
   (a) Increase with altitude
   (b) Decrease with altitude
   (c) First decreases and then becomes constant
   (d) Remains constant.

180. Which of the following remain constant in a planetary motion (consider elliptical orbits) or seen from the sun?
   (a) Speed (b) Angular speed
   (c) Kinetic energy (d) Angular momentum.

181. A cosmic year is equal to solar year
   (a) 250 million (b) 450 million
   (c) 4500 million (d) 600 million.

182. An Altimeter is used to measure?
   (a) Height above earth surface
   (b) Depth below sea level

(c) Length of mountain
(d) None of these.

183. By which force we seperate cream from milk?
(a) Gravitation force (b) Centripetal force
(c) Centrifugal force (d) Molecular force.

184. To find magnified image from a concave mirror of focal length 20 cms at what distance from the mirror should one stand ?
(a) 0-10 cm (b) 10-10 cm
(c) 20-40 cm (d) 40-80 cm.

185. Name the planet of the solar system that revolves slowest around the sun
(a) Pluto (b) Venus
(c) Mars (d) Mercury.

186. Which instrument is used for measuring relative humidity of air?
(a) Barometer (b) Manometer
(c) Lactometer (d) Hygrometer.

187. At what temperature do the Fareinheit and Centigrade thermometers show the same reading?
(a) -40° F (b) 0° F
(c) °C (d) 100F.

188. Which type of lense is used for viewing objects coming from behind in a motor car?
(a) Convex lens (b) Concave lens
(c) Plain glass (d) None of these.

189. By whom was the Barometer invented?
(a) Bertholet (b) Bragg
(c) Torricelli (d) Brand.

190. What for is a barometer used?
(a) To measure atmospheric pressure
(b) To measure gas

(c) To measure height
(d) To measure depth.

191. Name the unit of power?
(a) Volt (b) Ampere
(c) Watt (d) Kilowatt.

192. What change in the volume of water will you observe if you cool it from 4°c to 0°c?
(a) Increase (b) Decrease
(c) No change (d) None of these.

193. In which form does the earth receive the largest amount of energy from the Sun?
(a) IR (b) U.V
(c) Visible light (d) X-rays.

194. Name the instrument that you use to measure the diameter of a thin wire.
(a) Screw guage (b) Drapes
(c) Vernier callipers (d) None of these.

195. The colour of sky as observed by an astronaut in space is
(a) Blue (b) Black
(c) Colourless (d) White.

196. By whom was the quantom theory developed
(a) Ferah (b) Plank
(c) Bohr (d) Thomson.

197. By whom was the law of gravitation propunded ?
(a) Newton (b) Summerfield
(c) Nicholas (d) Grant.

198. What was Galileo's first scientific descovery?
(a) Telescope (b) Pendulum
(c) Gravity (d) Gravitation.

199. By whom was the microscope invented ?
(a) Hock (b) Brook
(c) Henry (d) Soddy

200. How much time does the planet mercury take to complete one revolution round the sun?

(a) 24 hours (b) 88 days
(c) 365 days (d) 180 days.

201. Who proved that when electric current passes through a conductor it produces a magnetic field?

(a) Oersted (b) Ohm
(c) Grahm Bell (d) Thomson.

202. A physical constant has been named after the physicist who propounded the quantum theory who was he?

(a) Rydberg (b) Plank
(c) Universal (d) Paschen.

203. What far an electroscope is used?

(a) Electric charge (b) Magnetic field
(c) Gravitation field (d) Matter.

204. Name the material that is used to make the cores of a transformer

(a) steel (b) stainless steel
(c) soft iron (d) None of these.

205. At what height is a geostationary satellite situated?

(a) 10,000 km (b) 36,000 Km
(c) 3600 km (d) 1000 km.

206. At what temperature does water have minimum specific heat?

(a) 0°c (b) 100°c
(c) 31°c (d) None of these.

207. Is gravitational effect associated with electric current?

(a) Yes (b) No
(c) Not sure (d) Can not be predicted.

208. What is the unit of loudness?

(a) Mach (b) Phon
(c) Mic son (d) none of these.

209. How is the frequency of tuning fork affected by increase of temperature?
(a) Increases (b) Decreases
(c) Remains same (d) Can't say.

210. What for are electrometers used?
(a) To measure (b) Charge
(c) Current (d) charge density.

211. Which one inductance or capacitance a resister possesses ?
(a) Inductance (b) Capacitance
(c) Both (d) None.

212. For a tube which own more? Its D.C resistance or its AC resistance?
(a) A.C. (b) D.C.
(c) Same for both (d) None of these.

213. As the atomic number increases electron affinity
(a) Decreases
(b) Increases
(c) Remains constant
(d) Approaches infinite value.

214. Name the waves that are used by Radar to detect an object
(a) I.R. waves (b) Micro waves
(c) Radio waves (d) None of these.

215. Give the name of the instrument that is used by sailors to see objects hidden from direct view
(a) Kaleiodscope (b) Periscope
(c) Telescope (d) Microscope.

216. One Faraday is equal to ---------------- culombs.
(a) 1 (b) 96540
(c) 48270 (d) 1000.

217. What for an ammeter is used?
(a) Potential (b) Current strength
(c) Power (d) None of these.

218. Name the instrument that is used for measuring the altitude of the sun

(a) Altimeter (b) Barometer
(c) Sextant (d) Manometer.

219. Name the unit to measure the luminous intensity of a source light

(a) Ampere (b) Photon
(c) Einsten (d) Candela.

220. Name the instrument used for comparing the luminous intensities of two sources of light:

(a) Viscometer (b) Stalgometer
(c) Photometer (d) Ammeter.

221. Who pioneered Radio astronomy?

(a) Rontegen (b) Mark Twain
(c) Karl Marx (d) Karl Jansky.

222. Laser was invented in 1960 by an American scientist. Name the scientist?

(a) J. Maiman (b) Hann
(c) Herst (d) Highs.

223. Expand the word LASER.

(a) Liquid adjusted reaserch
(b) Light amplification by stimulated emission of radiation
(c) Light and sea energy reaserch
(d) Land, sun, earth, revolution.

224. Name the place where the headquarters of World Meteological Organisation are situated

(a) Chicago (b) Geneva
(c) New York (d) Washington.

225. Name the only planet that was discovered in the 20th century.

(a) Uranus (b) Neptune
(c) Pluto (d) Venus.

226. How is match expressed in miles/hours?

(a) 7960 miles/hours (b) 7690 miles/hours

(c) 9760 miles/hours (d) 9670 miles/hours.

227. By whom was the first electronic motor made?

(a) Write Brothers (b) Graham Bell

(c) Michael Faraday (d) Einstein.

228. How is nautical mile related to Kilometers?

(a) 1.85 km (b) 1.25 km

(c) 0.95 km (d) 0.25 km.

229. When and by whom was the first printing press developed?

(a) In 1943, Gultenberg (b) In 1940, Thomson

(c) In 1950, Oxford (d) In 1900, Printfard.

230. When was the first telescope made?

(a) 1600 AD (b) 1500 AD

(c) 1700 AD (d) 1800 AD.

231. By whom was the first reflecting telescope made?

(a) Plato (b) Aristotle

(c) Newton (d) Raman.

232. The first steem engine was launched by George Stephenson in 1825. What was it called?

(a) Steam engine (b) Train

(c) Locomotive (d) Motor.

233. Name the scientist who first proved that lighting is an electric discharge coming down to the earth.

(a) Hamprhire (b) Thomson

(c) Newton (d) Benjamin Franklin.

234. Why does the sky look blue?

(a) Blue colour of sky

(b) Blue colour present in sunlight

(c) Scattering of light

(d) Variation in light.

235. Name the scale that is used to measure the intensity of earthquakes.

(a) Ritche (b) Pitcher
(c) Eicher (d) Mach.

236. Name the person who built the first radio telescope.

(a) Gallelio (b) Aristotle
(c) Jansky (d) Newton.

237. By whom was the radio invented?

(a) Grahm Bell (b) Marconi
(c) Thomson (d) Theopord.

238. What is the velocity of light?

(a) 2×10^{10} cms (b) 3×10^{8} ms
(c) Both (a) and (b) (d) 3×10^{6} ms.

239. By whom was lightening conductor invented?

(a) Faraday (b) Franklin
(c) Frankland (d) Ford.

240. The solar system comprises of

(a) 9 planets
(b) 9 planets and 37 satellites
(c) 9 planets, 31 satellites and countless asterads, comets etc.
(d) 9 planets 37 satellites and countless asterolds.

241. Stars would be seen clearly

(a) If we cross the earth
(b) If we cross the sky
(c) If we cross the earth and sky
(d) If we cross the atmosphere

242. The planet to the sun is

(a) Earth (b) Venus
(c) Mercury (d) Jupiter.

243. Which of the following is not present in solar system

(a) Venus (b) Jupiter
(c) Pluto (d) Air.

244. The ıargest planet in the solar system is
(a) Pluto (b) Saturn
(c) Earth (d) Jupiter.

245. Meteors are small bites of
(a) Soft iron (b) Mineral matter
(c) Hard matter (d) None of the above.

246. Meteors are
(a) Weight from a few ounces to even nons
(b) Weight from one kg to 10 kg
(c) Weightless
(d) All are correct.

247. The planet farthest away from the Sun is
(a) Uranus (b) Pluto
(c) Saturn (d) Uranus.

248. The path of a comet is so long that we may see it, once only in
(a) Sixty years (b) Hundred years
(c) Fifty years (d) None of the above.

249. A comet moves around the
(a) Earth (b) Moon
(c) Sun (d) Sky.

250. The largest planet in the solar system is
(a) Pluto (b) Jupiter
(c) Earth (d) Venus.

251. A comet looks like a
(a) Ball
(b) Earth
(c) Animal body with long tail
(d) Star with a long tail.

252. The planet called the Twin of the earth is
(a) Jupiter (b) Pluto
(c) Venus (d) Saturn.

253. The temperature of the solar surface is about

(a) 5000°C (b) 6000°C

(c) 7000°C (d) 8000°C.

254. Which of the following statements about the sun is wrong

(a) Rounded in structure

(b) It gives light

(c) Size of sun is more than earth

(d) It completes one rotation on its axis in 30 days.

255. A dry cell is just like a

(a) Deniel cell (b) Voltaic cell

(c) Leclanche cell (d) None of these.

256. The angel of contanct between pure water and pure glass

(a) 0° (b) 45°

(c) 90° (d) 135°.

257. Radiation emitted by a surface is directly proportional to

(a) Third power of its temperature

(b) Fourth power of its temperature

(c) Twice power of its temperature

(d) Equal to its temperature.

258. Which of the following is an example of scalar quantity?

(a) Velocity (b) Force

(c) Angular momentum (d) Electrostatic potential.

259. The frictional force is

(a) Self-adjustable (b) Not self-adjustable

(c) Vector quantity (d) Scalar quantity.

260. How many images will be formed if two mirrors are fitted on adjacent wall and one mirror on ceiling ?

(a) 2 (b) 5

(c) 7 (d) 11.

261. Magnetic permeability is maximum for

(a) Diamagnetic substance

(b) Paramagnetic

(c) Ferromagnetic substances
(d) All of these.

262. Which of the following not deflected by a magnetic field when it is in motion?
(a) Protons
(b) Neutrons
(c) Alpha-particles
(d) Beta-particles.

263. If earth stand then weight of a man will
(a) Increase (b) Decrease
(c) Remain the same (d) Become zero.

264. The pitch of note depends upon its
(a) Frequency (b) Savelength
(c) Amplitude (d) Speed.

265. Which of the folowing statements is false ?
(a) Mass, speed and energy are scalars
(b) Momentum, force and torque are vectors
(c) While distance is a scalar and displacements is a vector
(d) A vector has only magnitude whereas a scalar has both magnitude and direction.

266. A body travels such that its displacements is proportional to the square of the time taken its acceleration is
(a) Zero (b) Infinite
(c) Constant (d) Variable.

267. The colour of light which is least deviated on passing through a glass prism is
(a) Red (b) Blue
(c) Violet (d) Yellow.

268. Decibel is a
(a) Musical note
(b) Harsh noise
(c) Musical instrument
(d) Unit of measuring loudness of sound.

269. The Freezer in a refrigerator is fitted near the top
   (a) To keep it away from the hot compressor which is near the bottom
   (b) Because of convenience
   (c) So that it can cool the whole interior by setting up convection currents
   (d) Without any specific purpose.

270. One feels cold after a dip in sea water due to
   (a) Convection currents
   (b) The disposition of salt on the body
   (c) The evaporation of water on the body
   (d) The difference in temperatures between land and sea.

271. The walls and ceiling of an auditorium are covered with fibrous material such as glass fibres to
   (a) Beutify the auditorium
   (b) Reduce the cost of construction
   (c) Make the auditorium fire-proof
   (d) Absorb sound and prevent echoes.

272. Air is completely removed from an electric bulb to prevent
   (a) Oxidation of tungsten filament
   (b) Bursting of the bulb
   (c) Loss of light due to absorption
   (d) None of these.

273. The process by which energy is generated in the sun is the
   (a) Fission of uranium
   (b) Fussion of helium
   (c) Fussion of hydrogen
   (d) Combination of all these processes.

274. Treads on the shoe soles are made to increase
   (a) Strength (b) Durability
   (c) Friction (d) Elasticity.

275. Hydrogenic brakes in automobile work on
   (a) Bernoulli's principle (b) Posieuilles principle
   (c) Pascal's principle (d) Archimedes principle.

276. After a shower of rain a rainbow is seen
   (a) Towards the sun
   (b) Opposite
   (c) Anywhere irrespective of the position of the sun
   (d) Even in the absence of the sun.

277. On a clean glass plate a drop of water spreads to form a thin layer where as a drop mercury remains almost spherical because
   (a) Mercury is a metal
   (b) Density of mercury is greater than that of water
   (c) Cohesion of mercury is greater than its adhesion with glass
   (d) Cohesion of water is greater than its adhesion with glass.

278. Which one the following would a hydrogen ballon find easiest to lift
   (a) 1 Kg of water
   (b) 1 Kg of copper
   (c) 15 Kg of loosely packed fathers
   (d) All of the abvove.

279. In a cup tea cools from 90°C to 80°C in exactly one minute. It will cool from 70°C to 60°C in
   (a) Less than one minute
   (b) Exactly one minute
   (c) Nearly one minute
   (d) More than one minute.

280. The surface of water in a pond remains cool even on a hot day because
   (a) Water evaporates at the surface and evaporation causes cooling
   (b) The heat absorbed at the surface is transmitted downwards
   (c) Water in contact with the bottom of the pond becomes cool and rises to the surface due to convection
   (d) None of these.

281. The types of mirror used in the headlamp of cars is the
   (a) Plane mirror
   (b) Spherical convex mirror
   (c) Spherical concave mirror
   (d) Parabolic.

282. The starting current of an electric motor is .... the current flowing after the motor reaches its normal running speed
   (a) Nearby equal to (b) Exactly equal to
   (c) Less than (d) Much greater than.

283. Energy conversions taking place microphone are
   (a) Sound to mechanical to electrical
   (b) Electrical to mechanical to sound
   (c) Sound to mechanical
   (d) Mechanical to sound.

284. The process responsible for the enormous energy in the sun is
   (a) Nuclear fission (b) Nuclear fusion
   (c) The burning of gases (d) As yet unknown.

285. The inside of a photographic camera is coated black to avoid
   (a) Scattering of light
   (b) Total internal reflection of light
   (c) Reflection of light
   (d) Absorption of light.

ANSWER SHEET

1.	(b)	2.	(a)	3.	(d)	4.	(d)
5.	(b)	6.	(c)	7.	(d)	8.	(b)
9.	(d)	10.	(c)	11.	(a)	12.	(d)
13.	(a)	14.	(c)	15.	(b)	16.	(b)
17.	(b)	18.	(b)	19.	(a)	20.	(c)
21.	(a)	22.	(a)	23.	(b)	24.	(c)
25.	(c)	26.	(c)	27.	(a)	28.	(c)
29.	(b)	30.	(a)	31.	(b)	32.	(c)

33.	(b)	34.	(b)	35.	(a)	36.	(b)
37.	(b)	38.	(a)	39.	(d)	40.	(a)
41.	(b)	42.	(b)	43.	(d)	44.	(d)
45.	(b)	46.	(a)	47.	(b)	48.	(d)
49.	(b)	50.	(a)	51.	(d)	52.	(d)
53.	(d)	54.	(b)	55.	(a)	56.	(b)
57.	(c)	58.	(a)	59.	(b)	60.	(b)
61.	(b)	62.	(b)	63.	(d)	64.	(a)
65.	(c)	66.	(d)	67.	(d)	68.	(a)
69.	(d)	70.	(a)	71.	(a)	72.	(a)
73.	(a)	74.	(a)	75.	(d)	76.	(d)
77.	(d)	78.	(d)	79.	(c)	80.	(b)
81.	(b)	82.	(c)	83.	(d)	84.	(b)
85.	(b)	86.	(a)	87.	(c)	88.	(b)
89.	(b)	90.	(a)	91.	(b)	92.	(a)
93.	(d)	94.	(b)	95.	(a)	96.	(c)
97.	(b)	98.	(c)	99.	(b)	100.	(d)
101.	(c)	102.	(a)	103.	(d)	104.	(c)
105.	(b)	106.	(c)	107.	(a)	108.	(c)
109.	(c)	110.	(c)	111.	(c)	112.	(a)
113.	(a)	114.	(b)	115.	(a)	116.	(a)
117.	(a)	118.	(a)	119.	(d)	120.	(d)
121.	(b)	122.	(b)	123.	(d)	124.	(b)
125.	(b)	126.	(d)	127.	(a)	128.	(d)
129.	(b)	130.	(a)	131.	(d)	132.	(a)
133.	(b)	134.	(b)	135.	(c)	136.	(d)
137.	(a)	138.	(c)	139.	(d)	140.	(d)
141.	(a)	142.	(a)	143.	(b)	144.	(c)
145.	(d)	146.	(a)	147.	(b)	148.	(a)
149.	(a)	150.	(d)	151.	(d)	152.	(d)
153.	(d)	154.	(c)	155.	(a)	156.	(c)
157.	(d)	158.	(c)	159.	(d)	160.	(a)
161.	(a)	162.	(b)	163.	(a)	164.	(b)
165.	(a)	166.	(b)	167.	(a)	168.	(c)
169.	(c)	170.	(d)	171.	(c)	172.	(a)
173.	(b)	174.	(b)	175.	(b)	176.	(a)
177.	(b)	178.	(c)	179.	(b)	180.	(d)
181.	(a)	182.	(a)	183.	(c)	184.	(c)
185.	(a)	186.	(d)	187.	(a)	188.	(a)
189.	(c)	190.	(a)	191.	(c)	192.	(a)
193.	(a)	194.	(a)	195.	(b)	196.	(b)

197.	(a)	198.	(b)	199.	(a)	200.	(b)
201.	(a)	202.	(b)	203.	(a)	204.	(c)
205.	(b)	206.	(c)	207.	(b)	208.	(b)
209.	(b)	210.	(a)	211.	(c)	212.	(c)
213.	(a)	214.	(c)	215.	(b)	216.	(b)
217.	(b)	218.	(c)	219.	(d)	220.	(c)
221.	(d)	222.	(a)	223.	(b)	224.	(b)
225.	(c)	226.	(a)	227.	(c)	228.	(a)
229.	(a)	230.	(a)	231.	(c)	232.	(c)
233.	(d)	234.	(c)	235.	(a)	236.	(c)
237.	(b)	238.	(a)	239.	(b)	240.	(c)
241.	(d)	242.	(b)	243.	(d)	244.	(a)
245.	(b)	246.	(a)	247.	(b)	248.	(c)
249.	(c)	250.	(b)	251.	(b)	252.	(b)
253.	(b)	254.	(d)	255.	(c)	256.	(a)
257.	(b)	258.	(d)	259.	(a)	260.	(c)
261.	(d)	262.	(b)	263.	(a)	264.	(a)
265.	(d)	266.	(c)	267.	(a)	268.	(d)
269.	(c)	270.	(c)	271.	(d)	272.	(a)
273.	(c)	274.	(c)	275.	(c)	276.	(b)
277.	(c)	278.	(c)	279.	(d)	280.	(a)
281.	(d)	282.	(d)	283.	(a)	284.	(b)
285.	(c)						

2
GENERAL ZOOLOGY

1. The science dealing with the study of human beauty is
   (a) Cryptology (b) Calology
   (c) Physiology (d) Psychology.
2. Which is the largest hairy animal?
   (a) Pythone (b) Lion
   (c) Blue whale (d) Elephant.
3. Graminivorous animals are
   (a) Grass eating animals (b) Flesh eating animals
   (c) New born animals (d) Laired animals.
4. Which is the carnivorous animals?
   (a) Cow (b) Rat
   (c) Lion (d) Elephant.
5. Carnivorous animals are
   (a) New born animals (b) Flesh eating animals
   (c) Haired animals (d) Grass eating animals.
6. Man is
   (a) A primate (b) A rodent
   (c) A mammal (d) An animal.
7. The animal which has highest length of the body in the world is
   (a) Diplocotr (b) Camel
   (c) Giraffe (d) Elephant.

8. The discovery of cholera germs was made by
   (a) Robert Koch (b) Ross
   (c) Louis (d) Bhava.

9. The theory of mutation was given by
   (a) Hugo de vries
   (b) The Darwin and Lamarck
   (c) Haldane
   (d) Fleming.

10. First chemical analysis of proportion was made by
    (a) Schleedan
    (b) Punet and Bateson
    (c) Mendel
    (d) Fleming.

11. Lamark had given theory of
    (a) Lamarkism
    (b) Isolation
    (c) Inheritance of aquired characters
    (d) Inheritance.

12. The origin of species was written by
    (a) Darwin (b) Colbest
    (c) Lamark (d) Mendel.

13. The father of Genetics is
    (a) Hon greger mendel (b) Lamark
    (c) Mendel (d) Bateson.

14. Who was the father of Biology?
    (a) Darwin (b) Mendel
    (c) Aristotle (d) Khurana.

15. Which is the organ in the following which does not take part in digestion?
    (a) Stomach (b) Liver
    (c) Kidney (d) Intestine.

16. The only vegetable proteins which is comparable in quantity to that of meat protein is
    (a) Ground nut (b) Soybean
    (c) Gram (d) Wheat.

17. A blanced diet is
    (a) Nutritionally adequate
    (b) Nutritionally inadequate
    (c) Sufficient to satisfy hunger
    (d) Rich in fats so as to produce adequate energy.

18. Proteins are mainly required for
    (a) Supply of calcium (b) Supply of energy
    (c) Supply of water (d) Regeneration of cells.

19. Proteins changed into amino acids by
    (a) Oxidation (b) Hydrolysis
    (c) Reduction (d) Decomposition.

20. Glucose is stored in the liver in the form of
    (a) Cellulose (b) Starch
    (c) Glycogen (d) Sucrose.

21. Which of the following statements is not correct?
    (a) Renin does not coagulate the milk
    (b) Ptyaline is found in saliva
    (c) Bile is secreted by liver
    (d) Starch is digested in mouth.

22. In alimentary canal starch is converted into
    (a) Fructose (b) Maltose
    (c) Glucose (d) Sucrose.

23. Glucose gives energy by
    (a) Hydrolysis (b) Oxidation
    (c) Reduction (d) Decomposition.

24. Iron deficiency causes
    (a) Anaemia (b) Cirronosis
    (c) Rickets (d) Goiter.

25. The disease beri-beri is caused by deficiency of vitamin
(a) A (b) D
(c) B (d) C.

26. The protein deficiency disease is known as
(a) Cirronosis (b) Kwashiorker
(c) Eczens (d) Hycoses.

27. Infertility is believed to be due to lack of vitamin
(a) B (b) A
(c) K (d) E.

28. Excess of vitamin D needs to
(a) rickets (b) Paralysis
(c) Osteomalria (d) Cirropsis.

29. The blood clotting factor is the vitamin
(a) K (b) A
(c) B (d) C.

30. The concept of vitamin as essential constituents of diets was given by
(a) Hook (b) Funk
(c) Alikunhi (d) Roomer.

31. The toxic substances in the diet are detoxified in human body by
(a) Heart (b) Liver
(c) Lungs (d) Kidney.

32. Enzymes are made of
(a) Fats (b) Protein
(c) Carbohydrates (d) Lipids.

33. The most abundant tissue in the body is
(a) Muscle (b) Nerves
(c) Epithelial (d) Vascular.

34. The most abundant protein in muscles is
(a) Mygolobin (b) Myosin
(c) Action (d) Actomyosin.

35. Which of the following is a non-calcified tissue?
    (a) Dentin (b) Bone
    (c) Liver (d) Enamel.

36. The animal acid which is metabolised in the brain is
    (a) Glutamic acid (b) Glysin
    (c) Aspotic acid (d) Leucine.

37. The number of pairs of spinal cord is
    (a) 30 (b) 31
    (c) 32 (d) 35.

38. Which of the following is not an organ of the central nervous system?
    (a) Vagins (b) Brain
    (c) Spinal cord (d) Carmial nerves.

39. The sweat glands do not occur in the dermis of
    (a) Eye lids
    (b) Lip margins
    (c) Anterior end of thumb
    (d) Nipples.

40. Two important mineral elements present in one matter are
    (a) Calcium and silicon
    (b) Phosphorus and silicon
    (c) Calcium and phosphorus
    (d) Calcium and alluminum.

41. Nurons are the cells that make the
    (a) Nerves (b) Heart
    (c) Kidney (d) Skin.

42. Lachcrymal glands secrete
    (a) Mucous (b) Saliva
    (c) Tear (d) Pus.

43. Trypsin is secreted by
    (a) Pancreas (b) Liver
    (c) Thyroid (d) Stomach.

44. Pancreatic juice digests both the
    (a) Carbohydrates and fat
    (b) Protein and carbohydrates
    (c) Fat and sugar
    (d) Protein and fat.

45. Fat soluble vitamins are
    (a) A, D and E (b) B and C
    (c) K & E (d) B_{12} and E.

46. Water soluble vitamins are
    (a) B and E (b) B and C
    (c) E and K (d) A, D and K.

47. Islets of Langerhans are found in
    (a) Thyroid (b) Uterus
    (c) Pancreas (d) Ovary.

48. Which of the following statements is wrong?
    (a) Bile juice is alkaline
    (b) Blood is acidic
    (c) Gastric juice contains hcl
    (d) Bile juice is secreted by lever.

49. After taking heavy diet one experience difficulty in breathing. This is because
    (a) Sternum does not move
    (b) Diaphragm can move less freely
    (c) The air passage is obsructed by the epiglottis
    (d) The movement of muscles becomes slow.

50. Which of the following statements is correct
    (a) The lizards has its long tail
    (b) Scurvy is a disease of bone
    (c) Tapole is the name of bone
    (d) Frown is the name of paloemon.

51. Four disease caused by virus are
    (a) Small pox all dyrrohea, dysentery, cholera

(b) Small pox, measils,fever, typhoid
(c) Influenza small pox measels foot and mouth diseases
(d) Nagana veginities, surra, influenza.

52. Deficiency of vitamin B_{12} causes
(a) Pellagra (b) Pernicious anaemia
(c) Rickets (d) Scurvy.

53. What is the name of the non-poisonous snake?
(a) Cobra (b) Naja
(c) Viper (d) Blind snake.

54. The flyless bird is
(a) Swift (b) Cuckoo
(c) Struthio (d) Parrots.

55. The mammal which lays eggs is
(a) Echidna (b) Mole
(c) Echidno (d) Marsupialia.

56. The amphioxus belongs to the
(a) Minor phylum (b) Invertebrate
(c) Vertebrate (d) None of the above.

57. Fish breathes with the help of
(a) Gills (b) Lungs
(c) Skin (d) Fins.

58. The starfish belongs to which phylum?
(a) Annelida (b) Mollusca
(c) Echinodermata (d) Pisces.

59. The first life on this earth came
(a) On land (b) On mountains
(c) In water (d) In air.

60. Which is the only mammal which can fly?
(a) Bat (b) Bee
(c) Snake (d) Whale.

61. The total number of muscles in the human body is
(a) 520 (b) 525
(c) 457 (d) 639.

62. Proteins are mainly required for
(a) Regeneration of cells (b) Supply of energy
(c) Supply of water (d) Supply of calcium.

63. The quantity of vanaspati ghee is improved by addition of vitamins
(a) B and C (b) A and B
(c) C and E (d) E and K.

64. A person doing light work should have a daily calorie in take of about
(a) 200 k cal (b) 2000 k cal
(c) 2500 k cal (d) 500 k cal.

65. The element which is required in largest amount by the human body is
(a) Calcium (b) Zinc
(c) Iron (d) Lead.

66. A man cannot survive without a certain amount of
(a) Water (b) Protein
(c) Carbohydrates (d) Fats.

67. Which is the richest source of carbohydrates among the following food items ?
(a) Potato (b) Meat
(c) Water (d) Vegetable.

68. Carbohydrates includes
(a) Starch and Water (b) Sugar and fats
(c) Starch and Sugar (d) Starch and Fats.

69. Which of the following is not an essential amino acid for human?
(a) Alanine (b) Lysine
(c) Valine (d) Tryptophan.

70. The essential fatty acids in the diet of many experimental animal is
    (a) Linoleic acid (b) Palmitic acid
    (c) Oleic acid (d) Stearic acid.

71. Which of the following constituents of food contains nitrogen ?
    (a) Fats (b) Proteins
    (c) Mineral salts (d) Carbohydrates.

72. Modern bread is fortified with the amino acid
    (a) Methiomine (b) Lysine
    (c) Leucine (d) Histidine.

73. Which of the following constituents of food contains carbohydrates ?
    (a) Amino acids
    (b) Alanine
    (c) Carbon containing compound
    (d) Nitrogen containing compound.

74. Food gives
    (a) Energy to the body
    (b) Mass in the body
    (c) Diet to alimentary canal
    (d) None of the above.

75. Influenza is caused by
    (a) Virus (b) Fungi
    (c) Algae (d) Bacteria.

76. Yellow fever is viral disease transmitted by
    (a) Culex (b) Aedes
    (c) Anopheles (d) All of them.

77. Filarial parasite is transmitted by
    (a) Male culex (b) Female culex
    (c) Male anopheles (d) Female anopheles.

78. Round worm is the comon name for
    (a) Nereis (b) Ascaris
    (c) Feretima (d) Taenia.

79. Which of the following disease cannot be prevented by vaccination?
    (a) Small pox (b) Poliomylitis
    (c) Cholera (d) Tuberculosis.

80. Which of the following is not a waterborn disease?
    (a) Asthma (b) Cholera
    (c) Amoebic dysentery (d) Typhoid fever.

81. Deptheria is caused by the bacterial infection of
    (a) Heat (b) Throat
    (c) Nasal passage (d) Lungs.

82. The tropical disease called kalazar is caused by
    (a) Leeshmania donovoni (b) Tse-tsefly
    (c) Entamoeba gingivalis (d) Entamoeba hystolytica.

83. The transmitting agents or vectors of malarial parasite are
    (a) Man
    (b) Female Anopheles Mosquito
    (c) Male Anoheles mosquito
    (d) Insects.

84. What is the name of the malarial parasite in man and other animals ?
    (a) Anopheles (b) Amoeba
    (c) Plasmodium (d) Moncystis.

85. Entamoeba gingivalis is found in the
    (a) Gums and teeth (b) Stomach
    (c) Intestine of men (d) None of the above.

86. Entamoeba hystolytica was first discovered by
    (a) Loseh (b) Craig
    (c) Dobell (d) Lambl.

87. Amoebic dysentery in man is caused by
    (a) Entamoeba hystolytica (b) Amoea
    (c) Paramecium (d) Endolimax.

88. Which of the following disease is caused by the bacteria ?
    (a) Dysentery (b) Cholera
    (c) Small pox (d) Influenza.

89. The disease small pox is caused by
    (a) Bacteria (b) Virus
    (c) Algae (d) Protozoa.

90. The acitivity of the thyroid gland is regulated by the pitutary harmone which is abbreviated as
    (a) ISH (b) TSH
    (c) ACTH (d) HCG.

91. An endocrine gland is one which secrets its product
    (a) To form chromation
    (b) To regulate transcription
    (c) Into a duct to be used locally
    (d) Into the blood stream to regulate the function of other organs.

92. Which of the following is not an endocrine gland?
    (a) Thyroid (b) Pancreas
    (c) Adrenol (d) Pituitary.

93. When fully expanded lungs of an average healthy person can hold volume of air corresponding to
    (a) 6 litres (b) 1.5 litres
    (c) 9 litres (d) 3 litres.

94. The hormone which regulates the rate of formation of red blood cells are secreted by
    (a) Liver (b) Pituitary
    (c) Kidney (d) Stomach.

95. The filtration of blood plasma is done by
    (a) Glomeruli (b) Loop of henle
    (c) Renal pelvis (d) Ureter.

96. The collecting chamber for the watery solution filtered from the blood in the tuft is

(a) Nephron (b) Renal tabule
(c) Bowman's capsule (d) Lumeular tuft.

97. The organ through which water is not lost is

(a) Lungs (b) Kidney
(c) Liver (d) Skin.

98. Which of the following is not a function of Kidney?

(a) Secretion of amebiotics
(b) Regulation of the acidity of body fluids
(c) Removal of urine
(d) Regulator of blood pressure.

99. Which of the following organs is not involved in excretion ?

(a) Kidney (b) Lungs
(c) Stomach (d) Skin.

100. Which of the following is not a gland?

(a) Liver (b) Kidney
(c) Pituitary (d) Pancreas.

101. The membranes enclosing the brain and spinal cord are known as

(a) Meninges (b) Antenal
(c) Nephron (d) Axon.

102. Study of the blood is known as

(a) Haematology (b) Plasmology
(c) Urology (d) Neurology.

103. The study of the nervous system and disorder is called

(a) Neurology (b) Herpetology
(c) Haematodology (d) Urology.

104. The weight of the human brain is about

(a) 700 g (b) 1800 g
(c) 1400 g (d) 2000g.

105. Total volume of blood in a normal man is
(a) 3 litres (b) 5-6 litres
(c) 7-9 litres (d) 10 litres.

106. Bile slats play an important role in the absorption of
(a) Fat (b) Vitamin
(c) Protein (d) Sugar.

107. The main organ for digestion and absorption of food is
(a) Stomach (b) Large intestine
(c) Small intestine (d) Liver.

108. The Bile is stored in
(a) Lungs (b) Gall bladder
(c) Liver (d) Spleen.

109. Bile is secreted by the
(a) Liver (b) Gall bladder
(c) Stomach (d) Small intestine.

110. The enzyme which coagulates the milk in man is
(a) Pepsin (b) Renin
(c) Trypsin (d) Lactose.

111. The enzyme Ptyaline is present in the
(a) Liver (b) Stomach
(c) Saliva (d) Intestine.

112. The length of the amintary canal is approximately
(a) 950 cm (b) 360 cm
(c) 760 cm (d) 260 cm.

113. The blood in human body is purified in
(a) The kidney (b) The heart
(c) The lungs (d) The veins.

114. The muscles of the heart are known as
(a) Striped muscles (b) Cordiac muscles
(c) Unstriped muscles (d) Smooth muscles.

115. The pericardium is the covering of the
(a) Lungs (b) Eye
(c) Heart (d) Kidney.

116. The bone radius is present in
(a) Human arm (b) Human griddle
(c) Human skull (d) Human leg.

117. The largest gland in human body is
(a) The pancreas (b) The liver
(c) The thyroid (d) The pituitary.

118. The number of vertebrae in human body is
(a) 13 (b) 23
(c) 33 (d) 43.

119. The pulmonary artery carries the
(a) Impure blood to lungs
(b) Pure blood to lungs
(c) Pure blood to the intestine
(d) None of the above.

120. The human heart is divided into
(a) Two chambers (b) Four chambers
(c) Five chambers (d) Three chambers.

121. Which area of the brain chamber possible for the control of the body temperature ?
(a) Thalamus (b) Pituitary
(c) Hypothelamus (d) Pineal.

122. The number of secral vertebrae which form sacrum in man is
(a) 1 (b) 2
(c) 3 (d) 5.

123. The skin on the palms and the sotes differs from the skin elsewhere in that
(a) Derms is extremely thick
(b) Skin in their region is a vascalor

(c) Epidermis is relatively thick with thicker keratimised outer layer

(d) None of the above.

124. Sweat glands and hair foilicles found in the dermis are developed from

(a) Cells of the epidermis
(b) Cells of the dermis
(c) Cells of the endodermis
(d) Cells of the hypodermis.

125. By what alternate name is the large feline carnivore known?

(a) Tiger (b) Snow leopard
(c) Chimpange (d) Stork.

126. Name the part of the human body in which inflamation is caused by the disease known as "encephalitis".

(a) Foot (b) Mouth
(c) Brain (d) Neck.

127. Name the dairy product in which vit.ellin is found.

(a) Milk (b) Egg
(c) Butter (d) Ghee.

128. Name the animal that lays the largest egg.

(a) Penguine (b) Shark
(c) Elephant (d) Camel.

129. Name the flightless bird that is known as the ostrich of south America.

(a) Shark (b) Rhea
(c) Doe (d) Elephant.

130. What is the colour of a silver fox's fur?

(a) Black (b) White
(c) Red (d) White frosted black.

131. The insect which lays eggs in water is

(a) Sandfly (b) Butterfly
(c) Dragonfly (d) Housefly.

132. Name the animals that migrate gross scandinavia and some times commit mass suicide in the sea.
(a) Hemmings (b) Flemingo
(c) Lemmings (d) Skylark.

133. Name the mammal that has the longest pregnancy in the world.
(a) Ostrich (b) Hemming
(c) Owl (d) Elephant.

134. What is the name given to the process in which an unfertilized egg can develop into an adult animal?
(a) Parthenogenesia (b) Life cycle
(c) Infancy (d) Nanatologesia.

135. Name the food of vampire bat
(a) Blood (b) Meat
(c) Plants (d) Eggs.

136. To which species glow-worms belong?
(a) Animals (b) Bird
(c) Betles (d) Reptiles.

137. The body temperature of warm blooded animals
(a) Changes with seasons
(b) Decreases in height
(c) Remains constant
(d) Increases in summer.

138. What is a Joey?
(a) Baby Jackal (b) Baby John
(c) Baby Elephant (d) Baby Pig.

139. The meat of an animal is called venison. Name the animal.
(a) Tiger (b) Oxen
(c) Pig (d) Deer.

140. Antigens are :
(a) Carbohydrate (b) Fat
(c) Hormone (d) Protein.

141. Entomology deals with the study of

(a) Animals (b) Birds
(c) Insects (d) Stars.

142. What is 'apodan'?

(a) Worm
(b) Animal without feet
(c) Both a and b
(d) None of the above.

143. What is the colour of a hippopotamus's perspiration?

(a) Colourless (b) White
(c) Black (d) Red.

144. How many pounds of grass can an adult hippopotamus's eat in one night?

(a) 100 pounds (b) 150 pounds
(c) 200 pounds (d) 250 pounds.

145. Can a camel store water in its hump?

(a) Yes (b) No
(c) Only in winter (d) Only in summer.

146. Enzyme ptyalin is produced by:

(a) Gastic gland (b) Islets of langerhans
(c) Liver (d) Salivary glands.

147. Name the Brazilian animal that will never live in its natural. habitat because the rain forests have been destroyed.

(a) Monkey (b) Ape
(c) Marmosets (d) Elephant.

148. Name the fastest member of the hare family.

(a) Antelope Jack Rabbit (b) Tiger
(c) Wolf (d) Leopard.

149. Which bird suspends itself in mid-air and feeds itself by thrusting its long toungue into flowers ?

(a) Humming bird (b) Owl
(c) Lemmings (d) Fowl.

150. Name the bird that covers maximum distance on its migratory journey.
(a) Arctic tern (b) Fleming
(c) Ostrich (d) Humming bird.

151. In which country the bird with the largest beak found?
(a) America (b) Africa
(c) Canada (d) England.

152. Who propounderd the theory of natural selection :
(a) Mendel (b) Lamark
(c) Darwin (d) De vries.

153. By what name is a male cat called
(a) Ton (b) Dick
(c) Harry (d) None of these.

154. Does the blood of an insect contains haemoglobin?
(a) Yes (b) No
(c) May be yes or no (d) None of the above.

155. How many toes an ostrich have on its each foot?
(a) 1 (b) 2
(c) 3 (d) 4.

156. Caral polyps build caral reefs. what are they?
(a) Plants (b) Animals
(c) Stones (d) Coal deposits.

157. Name the only mammal that flies
(a) Bat (b) Owl
(c) Ostrich (d) Fly.

158. A coenzyme is
(a) Nucleic acid (b) Organic compound
(c) Carbohydrate (d) Vitamin.

159. What is Bunny?
(a) Baby rabbit (b) Squirrel
(c) Baby rabbit or squirrel (d) None of these.

160. Give one word for the branch of science that deals with the study of the heart its physiology and pathology.

(a) Physiotherapy (b) Physiology
(c) Cardiology (d) Cardiogram.

161. What is a group of kangaroos called ?

(a) Mob (b) Hoe
(c) Doe (d) Dove.

162. Which bird is considered to be a symbol of love and peace?

(a) Nightiogram (b) Hans
(c) Dove (d) Parrot.

163. Give another name for termites

(a) Wood pecker (b) White ants
(c) Wood eater (d) None of these.

164. Name the one that is the rarest of all North American mammals

(a) Fern (b) Ferret
(c) Girraffe (d) Lagoons.

165. Name the tallest animal in the world.

(a) Giraffe (b) Elephant
(c) Ferret (d) Flemmingo.

166. Which of the following is the smallest functional genetic unit

(a) Chromosome (b) Cistron
(c) DNA (d) Gene.

167. What are Amphibians ?

(a) Vertebrates
(b) Non-vertebrates
(c) Cold blooded vertebrates
(d) Hot blooded vertebrates.

168. What are hermaphrodites?

(a) Animals that produce eggs
(b) Animals that produce sperms

(c) Animals that produce eggs and sperms
(d) Animals that produce neither egg nor sperms.

169. Name the animal that has instinctive urge to attack snakes.
(a) Mangoose (b) Lizard
(c) Moh (d) Owl.

170. Golgi body is found in:
(a) RBC
(b) All the cells
(c) All the cells excepts RBC and bacteria
(d) None of these.

171. The bird 'emu' is native of which country?
(a) England (b) America
(c) India (d) Australlia.

172. Name the most ancient living fossil vertebrate
(a) Latimeria (b) Mummy
(c) Wolfarm (d) None of these.

173. A wolfs bottle has .... necks ?
(a) 1 (b) 2
(c) 3 (d) Zero.

174. Which is the smallest dear?
(a) Pada (b) Pudy
(c) Ponda (d) Pome.

175. What does the joint and eater use as a blanket on cold nights?
(a) Its tail (b) Its claws
(c) Its fur (d) Its claws.

176. Name the animal that is called the red dog of Asia .
(a) Dog (b) Horse
(c) Dhole (d) None of these.

177. Name the rodent whose nick name is 'hazel' mouse.
(a) Rat (b) Darmouse
(c) Wild mouse (d) Red Rat.

178. Name the breed of animals to which shift zu1 vizsa and basenji belong.
   (a) Rabbit (b) Dogs
   (c) Horse (d) Cats.

179. The endoskeleton of the cell is made up of
   (a) Cytoplasm
   (b) Mitochondria
   (c) Galgi complex
   (d) Endoplasmic reticulum.

180. Name the reptiles that are the only surviving dinasarus.
   (a) Reptiles (b) Crocodiles
   (c) Lizzards (d) Elephants.

181. Name the prehistoric flying lizard that is considered to be the first back boned animal to fly.
   (a) Pterosaur or Archaeopteryx
   (b) Lambreda
   (c) Lambaster
   (d) None of these.

182. In an ant colony what is the sex of worker ants?
   (a) Male
   (b) Female
   (c) Some times males some times female
   (d) None of these.

183. Name the most destructive insect in the world.
   (a) Snake (b) Lizard
   (c) Locust (d) Enpheles.

184. Name the longest poisonous snake in the world.
   (a) Water snake (b) Desert snake
   (c) King cobra (d) Ordinary snake.

185. Name the bird which can fly under water.
   (a) Diver (b) Dipper
   (c) Dove (d) Dopler.

186. Which insect is responsible for transmitting the viral disease "African Horse Sickness or Equine plague"?
   (a) Floria worm (b) Glow worm
   (c) Bacile calmetle guarin (d) Culicodes.

187. Wich bacteria is used for BCG vaccine?
   (a) Baccile calmetle guorin (b) Typhoid
   (c) Antiviral (d) Toxin.

188. The viviparours animals are present in the phylum
   (a) Porifera (b) Amphibia
   (c) Mammalia (d) Protozoa.

189. Most trerrestrial animal reproduce by
   (a) Polliinarion (b) internal fertilization
   (c) Dichogamy (d) External fertilization.

190. The term gene was coined by
   (a) Johannsen (b) Barron
   (c) Darlington (d) Punneet.

191. Hereditary characters are passed from one generation to another through
   (a) Nucleoplasm (b) Protroplasm
   (c) Cytoplasm (d) Chromosomes.

192. Sex chromosomes may be defined as
   (a) X chromosmes
   (b) Y chromosome
   (c) Chromosome concerned with sex determination
   (d) SAT chromosome.

193. A sexual reproduction in paramecium is due to
   (a) Conjugation (b) Binary fission
   (c) Vegetation (d) Internal fertilization.

194. A sexual reproduction in protozoa is caused by
   (a) Conjugation (b) Fertilization
   (c) Binary fission (d) None of the above.

195. The genetic code was discovered by
(a) Khurana (b) Lamark
(c) Darwin (d) Gamow.

196. Which is the bisexual animal ?
(a) Earthworm (b) Pila
(c) Starfish (d) Amoeba.

197. Each tissue consists of cells which are
(a) Similar in origin and form and performing the same function
(b) Similar in origin and form but differing in function
(c) Similar in origin
(d) dissimilar in origin.

198. Testis and ovary are found in
(a) Plants (b) Animals
(c) Lower invertebrates (d) Higher vertebrates.

199. Growth is a
(a) Embryological process
(b) Virtual metalic process which brings a permanent change in size
(c) Neurological process
(d) None of the above.

200. Reproduction from an unfertilization egg may be defined as
(a) Apomixis (b) Plasmogamy
(c) Embrylogy (d) None of the above.

201. The variety of tissue using greatest amount of energy in a living system is the
(a) Connective tissue (b) Nervous tissue
(c) Epithelial tissue (d) Muscular tissue.

202. Reproductive cycle of animals and plants is characterised by the alternation of
(a) Zygote and spore formation
(b) Meiosis and fertilization

(c) Mitosis
(d) Metamorphosis.

203. The digestive vacuoles in the animal cells are
(a) Centrosomes (b) Ribosomes
(c) Mitochondria (d) Lysosomes.

204. The main controlling centre of the cells is the
(a) Mitochondria (b) Nucleus
(c) Ribosome (d) Endoplasmic reticulum.

205. The functional genetic unit is
(a) D.N.A. (b) Gene
(c) Chromosome (d) Cistrone.

206. Protoplasm of the nucleus is called
(a) Protoplast (b) Karyoplasm
(c) Cytoplasm (d) Chromoplasm.

207. The number of chromosome in a human cell is
(a) Antherlobes (b) Root tip
(c) Bark (d) Xylem.

208. Enzymatic synthesis of DNA molecule was first achieved by
(a) Watson (b) Darwin
(c) Kornberg (d) Khorane.

209. The genetic information is stored in
(a) R.N.A. (b) Ribosome
(c) D.N.A. (d) None of the above.

210. D.N.A differs from RNA by
(a) deoxyrybose and thymine present in D.N.A and uracile and ribose in R.N.A
(b) Thymine present in DNA and uracil in RNA
(c) Deoxyribose in RNA and ribose in DNA
(d) Deoxyribose in DNA and ribose in RNA.

211. Chromosomes can be stained specifically by
(a) Acetocarmine (b) Safranine
(c) Funusgreen (d) Amino black.

212. The molecular sructure of D.N.A. was proposed by
   (a) Mendel (b) Darwin
   (c) Khurana (d) Watson and Cricks.

213. The nucleic acids comprise
   (a) Deoxyribonucleic acid and ribonucleic acid
   (b) Dexoyribonuclic acid
   (c) Ribonucleic acid
   (d) D.N.A. and protein.

214. The enzymes and pigments associated with photosynthesis located in
   (a) Chromoplast (b) Chloroplast
   (c) Leucoplast (d) Plastic.

215. A.T.P synthesis takes place in
   (a) Protoplasm (b) Mitochondria
   (c) Lysosomes (d) Ribosomes.

216. The fundamental difference between the living and non-living pertains to
   (a) The existence of protoplasm
   (b) Interaction with oxygen
   (c) The property of reproduction
   (d) The property of locomation.

217. The cell theory was proposed by
   (a) Oken
   (b) Darwin
   (c) Hooke
   (d) Schleiann and Schleian.

218. In the cell nucleus takes part in
   (a) Cell division (b) Excretion
   (c) Digestion (d) Cell respiration.

219. Life may be defined as
   (a) Birth (b) Death
   (c) Birth and death (d) None of the above.

220. Virus represent the border line between the
   (a) Living and non-living (b) Life and death
   (c) Locomotion (d) None of the above.

221. The term "cell" was introduced by
   (a) Mendel (b) Darwin
   (c) Schleiden (d) Hooke.

222. In the cell which of the following organs is known as powerhouse
   (a) Lysosomes (b) Mitochondria
   (c) Ribosomes (d) Chloroplast.

223. Bacteria single celled animals and plants are grouped together as
   (a) Protista (b) Protozoa
   (c) Protophyta (d) Porifera.

224. Pre-school group children need diet rich in
   (a) Fat (b) Vitamins
   (c) Proteins (d) Carbohydrates.

225. Living objects need food to obtain
   (a) Water (b) Oxygen
   (c) Organic (d) Energy.

226. Silver fish is
   (a) An insect (b) A fish
   (c) A term used in politics (d) A fish made of silver.

227. Red blood corpuseeles are found in
   (a) Bone marrow (b) Heart
   (c) Liver (d) Spleen.

228. Fundamental properties of living matter are
   (a) Mobility
   (b) Mobility and growth
   (c) Mobility and replication
   (d) Synthesis growth and replication.

229. Which is the correct statement?
   (a) Whales are cold-blooded animals
   (b) Whales are mammals
   (c) Whales are biggest fish
   (d) Whlaes live in fresh water.

230. Lack of iron causes
   (a) Diptheria (b) Anaemia
   (c) Pyorrhea (d) Pneumonia.

231. Function of skin is
   (a) To protect the hair (b) To protect the veins
   (c) To protect the body (d) Protect blood vessels.

232. Which is the wrong statement
   (a) Skin is not part of excretion
   (b) Heart's function is only to keep the blood in motion
   (c) Lungs are situated on either side of the heart
   (d) None of these.

233. Which disease is caused by a virus?
   (a) Malaria (b) Cholera
   (c) Small pox (d) Typhoid.

234. Which of the following crops can be grown on tropical marshy as well as low lying areas
   (a) Wheat (b) Rice
   (c) Potato/Oat (d) Barley.

235. Non-functioning of kidney is due to
   (a) Accumulation of nitorgenous waste
   (b) Presence of water in blood
   (c) Low blood pressure
   (d) High blood pressure.

236. Percentage of water in human body is approximately
   (a) 20% (b) 30%
   (c) 40% (d) 65%.

237. Silkworm the silk insect eats
   (a) Barks (b) Roots
   (c) Mulbery leaves (d) Borks.

238. Which of the following is present in greater proportion in bones
   (a) Calcium (b) Iodine
   (c) Iron (d) Phosphorus.

239. What element is present in maximum amount in human body
   (a) Carbon (b) Oxygen
   (c) Phosphorus (d) Nitrogen.

240. Spider spins its web from a liquid secreted from its
   (a) Posterior abdomen glands
   (b) Mouth
   (c) Salivary glands
   (d) Posterior silk gland.

241. Penicilin an important antibiotics was invented by
   (a) Jonas salk (b) Alexander Fleming
   (c) Robbert Koch (d) Edward Jennen.

242. Largest number of atoms at which element constitute human body ?
   (a) Hydrogen (b) Oxygen
   (c) Nitrogen (d) Phosphorous.

243. Which part of the human body coordinates the various functions
   (a) Heart (b) Liver
   (c) Nerves (d) Brain.

244. The prabuse respiration due to
   (a) Hot waether (b) High humidity
   (c) Still wind (d) All of the above.

245. The energy of food is measured in
   (a) Bushel (b) Erg
   (c) Calorie (d) Kelvin.

246. Which is the main source of vitamin 'A'
(a) Lemon (b) Meat
(c) Green vegetables (d) Oranges.

247. The deficiency ot vitamin 'A' cuases
(a) Dyorrhea (b) Beri beri
(c) Night blindness (d) None of these.

248. A person suffering from diabetes can be cured by injecting
(a) Insulin (b) Penicillin
(c) Streptomycin (d) Antibiotics.

249. The sum total of chemical processes in the body is called
(a) Vernalisation (b) Photosynthesis
(c) Respiration (d) Metabolism.

250. What is the major component of bones and teeth
(a) Calcium nitrate (b) Calcium Sulphate
(c) Calcium Phosphate (d) Calcium Carbonate.

251. The average heart beat per minute in the human body is
(a) 52 (b) 65
(c) 68 (d) 70.

252. The energy of food is measured in
(a) Calorie (b) Erg
(c) Bushel (d) Kelvin.

253. After how many days rash appears on body after the attack of measals
(a) One day (b) Two days
(c) Three days (d) Four days.

254. Decomposition of organic matter is due to
(a) Fungi (b) Virus
(c) Bacteria (d) None of these.

255. In India people suffer from anaemia due to the lack of
(a) Iodine (b) Calcium
(c) Potassium (d) Iron.

256. The content of water is greater than fat, the plasma is more than proteins, proteins are more than fats and fats are less than plasma, which is the major constitue of human body?
(a) Proteins (b) Water
(c) Fats (d) Plasma.

257. Dialysis is used for a patient suffering from
(a) Lung trouble (b) Liver trouble
(c) Kidney trouble (d) Bronchitis.

258. Which is the following cows give maximum milk yield
(a) Jersey (b) Sahiwal
(c) Holstein (d) Red Sindhi.

259. Pulse reading is done by doctors to find our
(a) Blood pressure (b) Heart beat
(c) Temperature (d) Respiration.

260. Descrease in white blood cells results is
(a) Increase in antigens (b) Increase in antibodies
(c) No change (d) Decrease in antibodies.

261. Vitamin responsible for anti-sterile activity is
(a) Vitamin E (b) Vitamin B
(c) Vitamin K (d) Vitamin C.

262. Which of the following is not immunised by 'Triple Antigen'?
(a) Diptheria (b) Typhoid
(c) Tetanus (d) Whooping cough.

263. Which of the following is the correct enunciation of the "Theory of Darwin"
(a) Different species do not arise by genetic mutation
(b) Weak and strong always maintain a fixed proportion
(c) Survival of the fittest and struggle for existence
(d) None of the these.

264. In India it is found recently that incidence of malaria is increasing because

(a) On account of increasing population it has become impossible to maintain cleanliness everywhere
(b) Of poverty in villages
(c) Of poor sanitary conditions
(d) Mosquitos have become DDT resistent.

265. The temperature of pasteurization is selected so as to
(a) Kill all bacteria
(b) Store it for long time without coagulation
(c) Kill all micro-organism ans other harmful bacteria
(d) Kill the micro-organism.

266. The term "Test Tube baby" implies
(a) Fertilization of ovum takes place in uterus but develops in the test tube
(b) Fertilization takes place in uterus and embyo develops in uterus
(c) Fertilization of ovum takes place in the test tube and develops in the test tube itself
(d) Fertilization of ovum takes in the test tube and develops in uterus.

267. The function of the liver is to
(a) Promote digestion of food
(b) Store glucose as glucogen
(c) Promote respiration
(d) None of these.

268. Carbohydrates, proteins and vitamins which are responsible for energy, growth and vitality are obtained from
(a) Milk, pulses and cereals
(b) Pulses, vegetables and cereals
(c) Milk, pulses and vegetables
(d) Cereals, milk and vegetables.

269. In high mountain regions bleeding through nose occurs because
(a) The pressure of the blood in the capillarios is higher then outside pressure

(b) Blood pressure decreases of high altitudes
(c) The pressure at high altitudes is greater than that at the plains
(d) The pressure at high altitudes is greater than that of the plains.

270. In summer man with excess respiration feels weak because of
(a) Loss of carbohydrates through evaporation
(b) Loss of salts through evaporation
(c) Loss of more water through evaporation
(d) All the factors mentioned above.

271. BCG vaccination to be given to a new born child
(a) Within six months (b) Within seven hours
(c) Within 48 hours (d) Immediately after birth.

272. "Heart Burn" refers to an uneasy burning sensation in the
(a) Heart (b) Liver
(c) Throat (d) Stomach.

273. The number of chromosome in human cell is
(a) 48 (b) 44
(c) 50 (d) 46.

274. Yellow fever is a viral disease transmitted by
(a) Acdes (b) Culex
(c) Anopheles (d) All of them.

275. The number of bones that form human skull is
(a) 27 (b) 36
(c) 25 (d) 28.

276. Which of the following is not a gland
(a) Liver (b) Kidney
(c) Pancreas (d) Pituitary.

277. Living objects need food to obtain
(a) Energy (b) Water
(c) Oxygen (d) Organic matter.

278. For a good health the relative humidity of a room should be about
   - (a) 30 %
   - (b) 70 %
   - (c) 100 %
   - (d) 50 %.

279. The main organ for digestion and absorption of food is
   - (a) Small intestine
   - (b) Liver
   - (c) Stomach
   - (d) Long intestine.

280. After hatching by hen the young chicks come out of eggs within
   - (a) One week
   - (b) Two weeks
   - (c) Three weeks
   - (d) Four weeks.

281. Mouth and foot disease in cattle are caused due to
   - (a) Fungi
   - (b) Bacteria
   - (c) Virus
   - (d) Pennicilium.

282. Green plants in sea are useful for respiration of fish because
   - (a) They give out oxygen
   - (b) They take oxygen and give out carbon dioxide
   - (c) They give out oxygen and carbon dioxide simultaneously
   - (d) They give out carbon dioxide.

283. The function of kidney is
   - (a) Store of but
   - (b) Reproduction
   - (c) Excretion
   - (d) Digestion.

284. In human body the blood is oxygenated and purified in the
   - (a) Heart
   - (b) Lungs
   - (c) Liver
   - (d) Kidney.

285. Oxygen is transported in the blood of human being by
   - (a) Plasma
   - (b) Haemoglobin
   - (c) Carbon dioxide
   - (d) Lymph.

286. The heart beat of a young man in approximately
   - (a) 70 times per minute
   - (b) 72 times per minute
   - (c) 78 per minute
   - (d) None of these

287. Green colour of leaves is due to the presence of

(a) Chloroform (b) Haemoglobin
(c) Chlorophyll (d) None of these.

288. The blood in human body is purified in

(a) Lungs (b) Heart
(c) Liver (d) None of these.

289. How much time is taken by sunlight to reach the earth from sun?

(a) 8 sec (b) 8 min
(c) 10 sec (d) 1 hour.

290. What is the ultimate source of energy ?

(a) Sun (b) Moon
(c) Wind (d) Sea.

291. It is difficult to cook pulse in the altitudinal hights due to

(a) Loss boiling point (b) More boiling point
(c) Air pressure (d) All of the above.

292. Who was the worlds first parachute jumper

(a) Hillary (b) Tenzing
(c) Shun (d) Valadamir.

293. Name the Only ape that can stand and walk upright

(a) Gibbon (b) Ape
(c) Kangaroo (d) Kappa.

294. What is a grospeak?

(a) Small Bird (b) Large Birds
(c) Type of book (d) Sea Animal.

295. Which one is a deeper sleep. Aestivation or hibernation?

(a) Aestivation (b) Hibernation
(c) Both are same (d) Can't be said.

296. Name the two ducts by which ova pass from the ovary to the uterus in female mammals.

(a) Testor (b) Uteros
(c) The fallopion tubes (d) Lymphs.

297. Name the gland that secrets the harmone cartisone hormone.

(a) Ducta (b) Ductelen gland
(c) Aelernal gland (d) Testes.

298. Name of the only bird which can fly back ward as well as forward.

(a) Hen (b) Peacock
(c) Penguine (d) Humming Bird.

299. Guenon is

(a) Monkey (b) Itama
(c) Ape (d) Bird.

300. How many species of the guinea pig are known till now?

(a) 4 (b) 6
(c) 10 (d) 1.

301. The gymire belongs to the family?

(a) Head hog (b) Rodent
(c) Bird (d) Plant.

302. Which hyena does not 'laugh'?

(a) Red hyena
(b) White hyena
(c) Black hyena
(d) Striped hyena.

303. Name the largest living reptile.

(a) Alligator (b) Crocodile
(c) Both A and B (d) Python.

304. Bile secreted by the liver gets stored in the

(a) Gall bladder (b) Duodenum canal
(c) Liver it self (d) Spleen.

305. Stomatology in the branch of medicine that deals with the study of --------------------?

(a) Stomach (b) Head
(c) Mouth disease (d) Lungs.

306. A south Sfrican surgeon Christian Barnard performed a seccessful operation in 1967 for the first time what was it called?

(a) Kidney transplant (b) Tumar removal
(c) Heart transplant (d) Skilled surgery

307. What is the Number of neck bones found in a Girraffe?

(a) 3 (b) 5
(c) 7 (d) 9.

308. Name the diet component that a camel stores in its hump.

(a) Carbohydrates (b) Protein
(c) Fat (d) Water.

309. What is the female of a rabbit, deer, antelope, hare or Kangaroo called?

(a) Hog (b) Pomina
(c) Doe (d) Tae.

310. An element is known as "Coagulation" factor IV" because of its role in multiple phases of blood coagulation. Name the element.

(a) Iron (b) Calcium
(c) Vitamin (d) Phosphorus.

311. Name the largest animal on land

(a) Indian elephant (b) Australian elephant
(c) Kangaroo (d) African elephant.

312. Name the largest animal in the world

(a) Elephant (b) Blue whale
(c) Scorpion (d) Snake.

313. Where are Penguins formed?

(a) Antarctica (b) South India
(c) South America (d) Europe.

314. What a male ass is called ?

(a) Jack (b) Jackal
(c) Jackie (d) Jacob.

315. From the following, choose the one that is as offspring of an elephant, whale, hippopotamous, moose, seal, deer.
(a) Whale (b) Hippopotomus
(c) Moose (d) Dear.

316. What kind of an animal is a Jacob?
(a) Donkey (b) Horse
(c) Ape (d) Jackal.

317. By whom was peniciline discovered?
(a) Franklin (b) Faraday
(c) Fleming (d) Ford.

318. Name the chief milk producing animal of Peru.
(a) Cow (b) Buffalo
(c) Iiu (d) llaena.

319. Which of the following is a genetic information:
(a) Carbohydrates (b) Faty acids
(c) Nucleic acids (d) Proteins.

320. Which of the following refers to a group of lions ?
(a) Pride (b) Bride
(c) Probe (d) Drobe.

321. Name the Great Britain's only poisoinous snake.
(a) King cobra (b) Adder
(c) Spider (d) Serpent.

322. What is the colour of a snow shoe hare's fur in summer?
(a) Brown (b) White
(c) Black (d) Orange.

323. Pathway of energy in an ecosystem is
(a) Cyclic (b) Unidirectional
(c) Web like (d) None of these.

324. Name the animal that looks like a miniature Kangaroo.
(a) Little kangaroo (b) Bandicoot
(c) Boot (d) Bross.

325. Antiserum is a serum containing :

(a) Antibodies (b) Antigens

(c) Bacteria (d) Leucocytes.

326. Starch and sugar together are formed as

(a) Carbohydrates (b) Fats

(c) Glucose (d) Proteins.

327. Name the sea animal that is considered to be a link between vertebrates and invertebrates

(a) Batla (b) Amphioxus

(c) Cotla (d) Bihu.

328. What are Ballen Whales?

(a) Small Whales (b) Large Whales

(c) Whales without teeth (d) Whales with sharptech.

329. Which of the following is not an indocrine gland?

(a) Adrenal (b) Pancreas

(c) Pituitary (d) Thyroid.

330. Name an Indian animal that hounds at the same speed as a red Kangaroo.

(a) Horse (b) Camel

(c) Rabbit (d) Rhinocerous.

331. What are the names of body coverings of Hedge Hogs?

(a) Spines (b) Quilts

(c) Hairs (d) Spin.

332. What is the desert fox of North Africa called?

(a) Clever fox (b) Felcon

(c) Clout (d) Fennec.

333. Capybara is a -------------------.

(a) Fish (b) Rodent

(c) Bird (d) Insect.

334. How many pairs of claws does a crab have ?

(a) 1 (b) 2

(c) 3 (d) 5.

335. Name the fastest land Mammal.
(a) Lion (b) Tiger
(c) Cheetah (d) Deer.

336. Name the most poisounous fish
(a) Stone fish (b) Jelly fish
(c) Catla (d) Small Fish.

337. Which birds are called Satbhai?
(a) Jungle babbler (b) Jungo
(c) Parakeet (d) Parrot.

338. By whom was the molecular structure of DNA propounded?
(a) Watson (b) Bose
(c) Watson and Crick (d) Crick.

339. B.C.G vaccine is used for immunisation against which disease?
(a) Cholera (b) T.B.
(c) Tetanus (d) Polio.

340. Oology is the branch of science that deals with ---------------.
(a) Birds (b) Sea Birds
(c) Birds eggs (d) Oester.

341. Meningities affects which part of the body ?
(a) Hand (b) Feet
(c) Brain (d) Heart.

342. Ligamens and fendons in the body are composed of
(a) Connective tissue
(b) Epithelial tissue
(c) Muscular tissue
(d) Skeletal tissue.

343. What is the normal temperature of human body?
(a) 100° F (b) 98.6°F
(c) 35°C (d) 98.6°C.

344. A large number of people suffer from a disease where in bronchidal tubes become narrow due to spasm of bronchial muscles and swelling of the mucous membrane. Name the disease.

(a) Bronchites (b) T.B.
(c) Asthama (d) Cough.

345. Name the chemical elements present in bones and teeth
(a) Calcium and phosphorus
(b) Calcium only
(c) Phoshphorus only
(d) Iron.

346. Name the Indian born scientist who has been awarded a Nobel prize in the field of medicines.

(a) Har Gobind Khurana (b) C.V. Raman
(c) Megh Nath Shaha (d) P.C. Bose.

347. Name the Vitamin that gets destroyed easily during cooking as also during storage ?

(a) A (b) B
(c) C (d) D.

348. Name the part of human body that coordinates the various functions of different organs of human body.

(a) Heart (b) Eyes
(c) Brain (d) Stomach.

349. Name the largest gland found in human body.

(a) Tumar (b) Endocrine gland
(c) Typhoid gland (d) Liver.

350. Name the largest bone of human body

(a) Leg bone (b) Back bone
(c) Femur (d) Jaw bone.

351. Name the shortest bone of human body

(a) Leg bone (b) Jaw bone
(c) Ear bone (d) Nose bone.

352. Name the element present in blood
(a) Iron (b) Magnesium
(c) Calcium (d) Iodine.

353. Name the biggest water snake
(a) Cobra (b) Adder
(c) Annoconda (d) King Cobra.

354. What does Cetelogy deal with ?
(a) Cattles (b) Whales
(c) Human beings (d) Pigs.

355. Name the substance that helps in digestion of milk.
(a) Calcium (b) Lactose
(c) Rennin (d) Vitamin.

356. What is the colour of ocean water?
(a) Colourless (b) White
(c) Blue (d) Black.

357. Name the person who was born in sixth century B.C. and is known as the father of plastic surgery.
(a) Seargent (b) Susruta
(c) Holms (d) Wholer.

358. Name the oldest living vertebrate.
(a) Hag Fish (b) Dinosour
(c) Apes (d) None of these.

359. Who is known as father of modern biology ?
(a) Aristotle (b) Hock
(c) Henry (d) None of these.

360. How many compartments are there in the stomach of a Cow?
(a) 1 (b) 4
(c) 3 (d) 2.

361. What are herbivorous animals ?
(a) Grass eating
(b) Flesh eating

(c) Eat both grass and flesh
(d) Remain in hibernation.

362. Name the heaviest flying bird
(a) Ostrich (b) Hen
(c) Bustard (d) Parakreet.

363. Name the scientist who was first to see the cell under the microscope
(a) Hock (b) Hooke
(c) Henry (d) Nicholson.

364. Give a suitable word to represent the outer portion of the cytoplasm of a cell
(a) Plasma (b) Platelets
(c) Cytoplasm (d) Ectoplasm.

365. Who discovered the four blood groups ?
(a) Leulick (b) Black
(c) Landsteener (d) Hock.

366. The germplasm theory is related to
(a) Mendel (b) Weisimann
(c) Darwin (d) Lamarck.

367. For what was Dr. Albert Szeni-g- Yorgyl awarded Nobel prize?
(a) Vitamin C (b) Polio vaccin
(c) Anti rabbies vaccine (d) T.B vaccine.

368. Name the basic building blocks of proteins
(a) Protein fibres (b) Nucleic acids
(c) Amino acids (d) Nucleotides.

369. Jaundice is a disease of -----------------?
(a) Eyes (b) Brain
(c) Liver (d) Heart.

370. Which one is present in antiserum ?
(a) Antigen (b) Antibodies
(c) Antiviral (d) None of these.

371. List the blood group that is considered as universal recipient :

(a) A (b) B
(c) AB (d) O.

372. Give the scintific name of Indian ape:

(a) Indian Ape (b) Hanuman
(c) Hylobates (d) Aerobates.

373. Give one word that is used for egg laying mammals

(a) Mammals (b) Birds
(c) Prototherians (d) Flying birds.

374. What is studied in Herpetology?

(a) Rabbits (b) Frog
(c) Reptiles (d) All of the above.

375. What is studied in Entomology?

(a) Insects (b) Rabbits
(c) Frog (d) Pig.

376. Inchyology is the study of

(a) Frogs (b) Fishes
(c) Birds (d) Insects.

377. What is the area of study of Osteology ?

(a) Bones (b) Skin
(c) Hair (d) Muscles.

378. What is the name of the large bone between the knee and foot of human beings ?

(a) Femur (b) Tibio
(c) Typhns (d) Ankle.

379. Parasite insects pulex irritans more commonly known as ?

(a) Human flea (b) Mosquitoes
(c) Human flea (d) Spider.

380. Germology is the study of

(a) birds (b) stamps
(c) old age (d) plants.

381. The muscles of the heart are

(a) Voluntary (b) Involuntary
(c) Stripped (d) Unstripped.

382. Which of the following acts as the transmitting agent in malaria ?

(a) Bacteria (b) Virus
(c) Mosquito (d) None of the above.

383. Red blood corpuscles are formed in the

(a) lungs (b) bone marrow
(c) kidneys (d) heart.

384. Vaccine of which one of the following diseases is administeres orally?

(a) Polio (b) Measles
(c) Small pox (d) Yellow fever.

385. Deficiency of calcium in the body cause which one of the following?

(a) Scurvy (b) Beriberi
(c) Anaemia (d) Rickets.

386. The sex of a child is determined by

(a) blood (b) mother
(c) father (d) both father and mother.

387. Deficiency of which one of the following vitamins hampers normal blood coagulation?

(a) B (b) B_{12}
(c) A (d) K.

388. Foot and mouth disease is associated with

(a) Cat (b) Cattle
(c) Fowl (d) None of the above.

389. Hepatitis is caused due to

(a) Uncooked food (b) Virus
(c) Body temperature (d) Drinking water.

390. Theory of Pangenesis was given by

(a) Darwin (b) Lamarck

(c) Hugo de vries (d) Operin.

ANSWER SHEET

1.	(b)	2.	(c)	3.	(a)	4.	(c)
5.	(b)	6.	(c)	7.	(a)	8.	(a)
9.	(a)	10.	(b)	11.	(c)	12.	(a)
13.	(a)	14.	(a)	15.	(c)	16.	(b)
17.	(a)	18.	(d)	19.	(b)	20.	(c)
21.	(a)	22.	(b)	23.	(b)	24.	(a)
25.	(c)	26.	(c)	27.	(b)	28.	(c)
29.	(a)	30.	(b)	31.	(b)	32.	(b)
33.	(a)	34.	(b)	35.	(c)	36.	(a)
37.	(b)	38.	(a)	39.	(b)	40.	(c)
41.	(a)	42.	(b)	43.	(a)	44.	(b)
45.	(a)	46.	(b)	47.	(c)	48.	(b)
49.	(b)	50.	(a)	51.	(c)	52.	(b)
53.	(b)	54.	(c)	55.	(a)	56.	(b)
57.	(a)	58.	(c)	59.	(c)	60.	(a)
61.	(b)	62.	(a)	63.	(b)	64.	(c)
65.	(a)	66.	(b)	67.	(a)	68.	(c)
69.	(a)	70.	(a)	71.	(b)	72.	(a)
73.	(c)	74.	(a)	75.	(d)	76.	(b)
77.	(a)	78.	(b)	79.	(c)	80.	(a)
81.	(b)	82.	(a)	83.	(b)	84.	(c)
85.	(a)	86.	(d)	87.	(a)	88.	(d)
89.	(b)	90.	(b)	91.	(d)	92.	(b)
93.	(a)	94.	(c)	95.	(a)	96.	(c)
97.	(c)	98.	(a)	99.	(c)	100.	(b)
101.	(a)	102.	(a)	103.	(a)	104.	(c)
105.	(b)	106.	(a)	107.	(c)	108.	(b)
109.	(a)	110.	(b)	111.	(c)	112.	(a)
113.	(b)	114.	(b)	115.	(c)	116.	(a)
117.	(b)	118.	(c)	119.	(a)	120.	(b)
121.	(c)	122.	(b)	123.	(c)	124.	(a)
125.	(b)	126.	(c)	127.	(b)	128.	(b)
129.	(b)	130.	(d)	131.	(c)	132.	(c)

133. (d)	134. (a)	135. (a)	136. (c)
137. (c)	138. (b)	139. (d)	140. (d)
141. (c)	142. (b)	143. (d)	144. (b)
145. (b)	146. (d)	147. (c)	148. (a)
149. (a)	150. (a)	151. (a)	152. (c)
153. (a)	154. (b)	155. (b)	156. (b)
157. (b)	158. (d)	159. (c)	160. (c)
161. (a)	162. (c)	163. (b)	164. (b)
165. (a)	166. (b)	167. (c)	168. (c)
169. (a)	170. (c)	171. (d)	172. (a)
173. (b)	174. (b)	175. (a)	176. (c)
177. (b)	178. (b)	179. (d)	180. (b)
181. (a)	182. (b)	183. (c)	184. (c)
185. (b)	186. (d)	187. (a)	188. (a)
189. (b)	190. (b)	191. (d)	192. (c)
193. (a)	194. (c)	195. (d)	196. (a)
197. (a)	198. (d)	199. (b)	200. (a)
201. (d)	202. (b)	203. (a)	204. (b)
205. (d)	206. (b)	207. (a)	208. (c)
209. (c)	210. (a)	211. (c)	212. (d)
213. (a)	214. (b)	215. (b)	216. (a)
217. (d)	218. (a)	219. (c)	220. (a)
221. (d)	222. (b)	223. (a)	224. (c)
225. (d)	226. (b)	227. (a)	228. (d)
229. (a)	230. (b)	231. (c)	232. (a)
233. (c)	234. (b)	235. (d)	236. (d)
237. (c)	238. (a)	239. (b)	240. (a)
241. (b)	242. (a)	243. (d)	244. (d)
245. (c)	246. (c)	247. (c)	248. (a)
249. (d)	250. (c)	251. (d)	252. (a)
253. (a)	254. (c)	255. (d)	256. (b)
257. (c)	258. (a)	259. (b)	260. (d)
261. (a)	262. (b)	263. (c)	264. (d)
265. (c)	266. (d)	267. (b)	268. (d)
269. (a)	270. (c)	271. (d)	272. (d)
273. (d)	274. (a)	275. (d)	276. (b)
277. (a)	278. (d)	279. (a)	280. (c)
281. (c)	282. (a)	283. (c)	284. (b)
285. (b)	286. (b)	287. (b)	288. (a)
289. (b)	290. (a)	291. (b)	292. (c)
293. (a)	294. (a)	295. (b)	296. (c)

297. (c)	298. (d)	299. (a)	300. (d)
301. (a)	302. (d)	303. (c)	304. (a)
305. (c)	306. (c)	307. (c)	308. (c)
309. (c)	310. (b)	311. (d)	312. (b)
313. (a)	314. (a)	315. (b)	316. (c)
317. (c)	318. (d)	319. (c)	320. (a)
321. (b)	322. (a)	323. (a)	324. (b)
325. (a)	326. (d)	327. (b)	328. (c)
329. (b)	330. (d)	331. (a)	332. (d)
333. (b)	334. (d)	335. (c)	336. (a)
337. (a)	338. (c)	339. (b)	340. (c)
341. (c)	342. (b)	343. (b)	344. (c)
345. (a)	346. (a)	347. (c)	348. (c)
349. (d)	350. (c)	351. (c)	352. (a)
353. (c)	354. (b)	355. (c)	356. (a)
357. (b)	358. (a)	359. (a)	360. (b)
361. (a)	362. (c)	363. (b)	364. (d)
365. (c)	366. (b)	367. (a)	368. (c)
369. (c)	370. (b)	371. (c)	372. (c)
373. (c)	374. (c)	375. (a)	376. (b)
377. (a)	378. (b)	379. (c)	380. (c)
381. (b)	382. (c)	383. (b)	384. (a)
385. (d)	386. (c)	387. (d)	388. (b)
389. (b)	390. (a)		

3
GENERAL BOTANY

1. Biology is the
   (a) Study of life as recorded by fossil
   (b) Study of uniformity behind diversity in living things
   (c) Study of all living organism
   (d) Study of growth and differentiation in living organisms.

2. Palaeabotany is the study of
   (a) Bones (b) Fossils
   (c) Primates (d) Birds.

3. Study of diseases of different plants is
   (a) Pathology (b) Plant physiology
   (c) Anatomy (d) Physicology.

4. Pendrology is the study of
   (a) Grasses (b) Flowers
   (c) Trees and shrubs (d) Trees.

5. Agrostology is cultivation of
   (a) Oil seeds (b) Crops
   (c) Gradsses (d) Fruits.

6. Genetics is study of
   (a) Forms and features of plant organs
   (b) Heredity and variations
   (c) Classification based character of somatic chromosomes
   (d) Growth and differentiation in living organisms.

7. Study of living activities and functions is called as
   (a) Cytology (b) Genetics
   (c) Ecology (d) Physiology.

8. The science of vegetable culture is called as
   (a) Floriculture (b) Agriculture
   (c) Otericulture (d) Horticulture.

9. Supermology deals with the study of
   (a) Seeds (b) Leaves
   (c) Fruits (d) Pollen grains.

10. Ethnobotany is a branch of Botany dealing with
    (a) Algae
    (b) Ether forming plants
    (c) Relationship between the primitive people and the plants
    (d) Myxomyceyts.

11. Study of light effects on living organisms constitute
    (a) Photobiology (b) Photomorphogenesis
    (c) Photoperiodism (d) Vernalisation.

12. Floriculture is the branch of
    (a) Horticulture (b) Botany
    (c) Agronomy (d) Entamology.

13. Pedology is the science of
    (a) Disease (b) Soil
    (c) Pollution (d) Rocks.

14. Algae and fungi are grouped in which of the following :
    (a) Thallophytes (b) Gymnosperms
    (c) Angiosperms (d) Bryophytes.

15. Which of the following plants have maximum power of adaptability?
    (a) Pteridophytes (b) Angiosperms
    (c) Gymnosperms (d) Bryophytes.

16. The basic unit of classification is
    (a) Genus (b) Species
    (c) Family (d) Order.
17. Which of following is known as father of taxonomy?
    (a) Engles (b) Aristotle
    (c) Linnaeus (d) Eichler.
18. National Botanical Garden is situated at
    (a) Chennai (b) Calcutta
    (c) Lucknow (d) Dehradun.
19. Central National Herbarium is situated at
    (a) Chennai (b) Sibpur (Calcutta)
    (c) Tarapur (Mumbai) (d) Dehradun.
20. Who is the father of Botany?
    (a) Theophrastus (b) Brunfels
    (c) Carolus Linnaeus (d) OswaldTippo.
21. One of the following events in a vegetable garden is never directly effected by light that is
    (a) Flowering (b) Fertilization
    (c) Food manufacturing (d) Seed germination.
22. "Origin of Species" was written by
    (a) Wallace (b) Darwin
    (c) Lamarck (d) Mendel.
23. The generic name for cotton plant is
    (a) Crotalaria (b) Raphanus
    (c) Nicatiana (d) Gossypium.
24. The approximate number of plants so far discovered is
    (a) 1,00,000 (b) 2,00,000
    (c) 2,50,000 (d) 3,00,000.
25. Genus is a groups of similar and related
    (a) Families (b) Orders
    (c) Species (d) Genera.

26. Virology is the study of
    (a) Bacteria (b) Virus
    (c) Algae (d) Fungi.

27. Microbiology deals with the study of
    (a) Animals
    (b) All plants
    (c) Microscopic plants
    (d) Microscopic plants and animals known as microbes.

28. The science which deals living organism is
    (a) Physics (b) Mathematics
    (c) Chemistry (d) Biology.

29. The science of naming of the plants is known as
    (a) Classification (b) Identification
    (c) Nomenclature (d) Taxonomy.

30. The name of the author of the species is always printed in
    (a) Roman type (b) Halic type
    (c) Bold type (d) Bold italic type.

31. The binomial system of nomenclature in plants was first used by
    (a) Darwin (b) Mendel
    (c) Lawrence (d) Linnaeus.

32. The branch of Botany dealing with heredity and variation is
    (a) Geobotany (b) Sericulture
    (c) Genetics (d) Evolution.

33. Paleontolgy is study of
    (a) Fossil plants (b) Fossil animals
    (c) Fossil grains (d) Both (a) and (b).

34. The branch of study of botany dealing with the study of Lichens is called
    (a) Mycophycology (b) Lichenology
    (c) Phycomycology (d) Microbiology.

35. Branch of Batany involved with isolation and fusion of protoplast is called
(a) Agriculture (b) Horticulture
(c) Tissue culture (d) Aqua culture.

36. Oncology is the study of
(a) Cell (b) Cancer
(c) Gene (d) DNA.

37. Study of nuclear cytoplasm is
(a) Genetics (b) Heredity
(c) Karyology (d) Cell biology.

38. Study of seasonal ornamental plants and flowers is known as
(a) Anthology (b) Agrostology
(c) Palynology (d) Phenology.

39. Palantology is the study of
(a) Fossils (b) Bones
(c) Primates (d) Birds.

40. Edaphology is the study of
(a) Embroyogenesis
(b) Pollen development
(c) Terrestrial vegetation
(d) Soil.

41. Prof. Birbal Sahni is known for his researches in the field of
(a) Palaeobotany (b) Physiology
(c) Cytology (d) Ecology.

42. Study of fossils plants is called
(a) Palaeontology (b) Ethonobotany
(c) Palaeobatany (d) Archaelogy.

43. Study of fruit science is termed
(a) Anthology (b) Pomology
(c) Palynology (d) Dendrology.

44. Who is the father of Indian plant physiology ?
   (a) Prof. Ranjan (b) Prof. Maheshwari
   (c) Sri J.C. Bose (d) Prof. Mehra.

45. Which is the recent branch of taxonomy ?
   (a) Karyotaxonomy (b) Cytotaxonomy
   (c) Morphotaxonomy (d) Taxidermy.

46. The science of naming the plant comes under
   (a) Classification (b) Identification
   (c) Numerology (d) Taxonomy.

47. Study of algae is called
   (a) Mycology (b) Phycology
   (c) Bacteriology (d) Oncology.

48. Study of flowers is called
   (a) Phenology (b) Anthology
   (c) Agrostology (d) Palynology.

49. Mycology is the study of
   (a) Mycoplasma (b) Fungi
   (c) Algae (d) All of these.

50. The study of mycoplasma comes under
   (a) Mycology (b) Virology
   (c) Bacteriology (d) Microbiology.

51. Formation of fruit without fertilization is known as
   (a) Agrofurnitary (b) Polyembryony
   (c) Parthenocarpy (d) Parthenogeros.

52. Cell membrane is
   (a) Permeable (b) Selectively permeable
   (c) Semi permeable (d) Impermeable.

53. Ecology deals with the relation between organiser and its
   (a) Habitat (b) habits
   (c) Environment (d) Life cycle.

54. Name the branch of science that deals with fruit and its cultivation.
    (a) Horticulture (b) Floriculture
    (c) Pomology (d) Agriculture.

55. The first person to see a cell under microscope was
    (a) Robert Hook (b) A.Vlevwenhock
    (c) T. Schwan (d) M. Schleiden.

56. Mushrooms lack green colour due to lack of
    (a) Chlorophyll (b) Haemoglobin
    (c) Leaves (d) Nitrogen.

57. What type of plants are sundew, bladderwart and pitcher plants?
    (a) Water plants (b) Desert plants
    (c) Mountain plants (d) Insectivorous plants.

58. By what means do bees communicate about newly found flowers with nectar?
    (a) Dance (b) Smell
    (c) Dance and smell (d) Do not communicate.

59. Name the world's smallest flower.
    (a) Kamal (b) Surajmukhi
    (c) Wolfia (d) Bougenvellia.

60. The science which deals with the study of viruses is called
    (a) Karylogy (b) Oncology
    (c) Microbiology (d) Bacteriology.

61. Another name of anotomy is
    (a) Plant morphology
    (b) Internal morphology
    (c) Demography
    (d) Histology.

62. Most of the bacterial names have been derived from
    (a) German (b) Greek
    (c) Latin (d) Spanish.

63. ICBN (International code of botanical nomenclature) was adopted in the year-
    (a) 1853 (b) 1953
    (c) 1964 (d) 1979.

64. Thallophytes having chloroplast are
    (a) Bacteria (b) Fungi
    (c) Algae (d) Mosses.

65. Bacteria were first discovered by
    (a) A.V. Leuwenhock (b) Robert Hooke
    (c) Rober Koch (d) Louis Pasteur.

66. The group of bacteria was raised to the status of division by
    (a) Smith and Wolf (b) Bold and Tippo
    (c) Pasteur and Koch (d) Wolf and Wolf.

67. Bacteria are considered to be plants because they
    (a) Cannot move
    (b) Have a rigid cell wall
    (c) Can multiply by fission
    (d) Are present every where.

68. Bacteria which are smallest in size are
    (a) Cocci (b) Vibrio
    (c) Bacilli (d) Spirilia.

69. One of the following is the smallest organism
    (a) Viruses (b) Bacteria
    (c) Mycoplasma (d) Yeast.

70. Which of the following bacteria fixes CO_2 in to carbohydrates ?
    (a) Rhodospirllium (b) Rhizobium
    (c) Azotobacter (d) Clostridium.

71. Which of the following is symbiotic bacterium ?
    (a) Nitrobactor (b) Nitrosomonas
    (c) Rhizobium (d) Clostridium.

72. The bacteria which in association with some plant roots fix atmoshpheric nitrogen are called
(a) Rhizobium (b) Salmonella
(c) Pseudomonas (d) Bacillus.

73. A cell wall material present in blue-green algae and bacteria is
(a) Pectin (b) Chitin
(c) Muramic acid (d) cellulose.

74. Cell wall in gram positive bacteria is composed of :
(a) Lipid and protein (b) Murein
(c) Proteins only (d) Cellulose and pectin.

75. Cell membrane of bacteria is made up of
(a) Lipid and protein (b) Protein and cellulose
(c) Chitin (d) Cellulose and lipid.

76. The flagella of bacteria are composed of
(a) Protein (b) Lipid
(c) Carbohydrates (d) Amide.

77. Which of the following is not a bacterial feature?
(a) Parasitism (b) Fermentation
(c) Plastids (d) Fixation of nitrogen.

78. A spiral bacterium is called
(a) Diplococcus (b) Bacillus
(c) Coccus (d) None of these.

79. Which of the following causes cholera?
(a) Spirilla (b) Cocci
(c) Commas (d) Bacilli.

80. Which of the following is a bacterial disease?
(a) Measles (b) Small pox
(c) Rabbies (d) Tuberculosis.

81. Waksman was a well known
(a) Phycologist (b) Bryologist
(c) Pteridologist (d) Soil Microbiologist.

82. The role of bacteria is not significant in the production of
    (a) Vitamin B-Complex (b) Coffee
    (c) Cheese (d) Baked bread.

83. Which are largest in number ?
    (a) Algae (b) Fungi
    (c) Bacteria (d) Insects.

84. Which of the following is the best source of bacteria ?
    (a) Water (b) Air
    (c) Soil (d) Human body.

85. Pure culture of bacteria was first obtained by
    (a) Koch (b) Pasteur
    (c) Leeuwenhock (d) Lister.

86. Bacteria are mostly
    (a) Autotrophic (b) hetrotrophic
    (c) Photosynthetic (d) Symbiotic.

87. Which of the following is not multicellular?
    (a) Oedogonium (b) Chara
    (c) Spirogyra (d) Chlamydomonas.

88. In chlamydomonas chlorophyll is found in
    (a) Pyrenoides (b) Grana
    (c) Stigma (d) Thylakoids.

89. In India "sleeping sickness" is due to the bite of
    (a) Sandfly (b) White fly
    (c) Tsetse fly (d) Fruit fly.

90. Locomotion in amoeba is by
    (a) Flagella (b) Pseudopodia
    (c) Setade (d) Cilia.

91. The term "protista" was given by
    (a) John ray (b). George leopold cuvier
    (c) Ernst Hackel (d) Carlolus Linnaeus.

92. Who was awarded Nobel prize in 1902 for his work on Malaria?

(a) Manson (b) Ronad Ross
(c) Graham (d) Laveran.

93. Slime moulds are

(a) Acellular (b) Cellular
(c) Cellular and acellular (d) Subcellular.

94. The cells walls of diatoms are impregnated with

(a) Calcium (b) Magnesium
(c) Silica (d) Iron.

95. Protozoans are also called

(a) Multicellular (b) Unicellular
(c) Noncellular (d) Acellular.

96. In algae, cell wall is made up of

(a) Chitin (b) Suberin
(c) Cellulose (d) Cutin.

97. Gametes produced after reduction division are termed as

(a) Meiogameters (b) Mitogameter
(c) Coenogameters (d) None of these.

98. A zygospore is

(a) Diploid (b) Haploid
(c) Triploid (d) Sporophytic.

99. Which of the following is not multicellular ?

(a) Oedogonium (b) Chara
(c) Spirogyra (d) Chlamydomonas.

100. Sexual reproduction in chlamydomonas is

(a) Isogamous (b) Anisogamous
(c) Oogamous (d) All of these.

101. Name the largest member of the grass family

(a) Green grass (b) Bamboo
(c) Coconut (d) Ashoka.

102. To which of the following funilised all. pulses belong ?
(a) Cruciferae (b) Leguminous
(c) Malvacead (d) Ranunculaceae.

103. Name the lightest wood in the world
(a) Balsa (b) Seesam
(c) Sagwan (d) Neem.

104. What are saparophytes?
(a) Producers (b) Decaying remains
(c) Scavangers (d) Pteridophytes.

105. Smallest flowering plant is
(a) Rafflesia (b) Polyalthia
(c) Wolfia (d) Chlamydomonas.

106. What type of plants grow in desert ?
(a) Hydrophytes (b) Xerophytes
(c) Ptederophytes (d) Sparophytes.

107. The food in onion is stored in the form of
(a) Cellulose (b) Protein
(c) Starch (d) Sugar.

108. Name the element present in chlorophyll.
(a) Iron (b) Magnesium
(c) Calcium (d) Iodine.

109. Which vegetation is known as "Pather ke phool"
(a) Lichen (b) Apple
(c) Guava (d) Grapes.

110. Who invented the crescograph?
(a) J.C. Bose (b) Subhash bose
(c) Bhabha (d) Brahmbhatta.

111. On which of the following plants did Greegar Mendel perform his classical experiments ?
(a) Gram (b) Maize
(c) Pea (d) Wheat.

112. Name two places where maximum number of fossils are found.
   (a) Utah and Holtz Maden
   (b) Andaman and Nicobar
   (c) Daman and Diu
   (d) Jerusulum and Zurich.

113. What is Hydrosphere?
   (a) Portion of earth (b) Forest
   (c) Sky (d) Atmosphere.

114. Edible part of Coconut is
   (a) Seed coat (b) Endocorp
   (c) Endosperm (d) Carpel wall.

115. Give the Botanical name of the cotton plant
   (a) Gossypium Hirsutum (b) Cotton Gispium
   (c) Both of these (d) None of these.

116. Name the process during which energy is released by plants
   (a) Photosynthesis (b) Respiration
   (c) Perspiration (d) Transpiration.

117. Give one word to indicate plants that give flower only once in their life time
   (a) Monocarpic (b) Monoloth
   (c) Monocot (d) None of these.

118. In which of the following the photosynthesis is maximum?
   (a) Green light (b) Blue light
   (c) Red light (d) Violet light.

119. Name the edible part of patato.
   (a) Root (b) Stem
   (c) Flavour (d) Fruit.

120. Mendel's law has been rediscovered in recent times. Name the person who did it.
   (a) Robbert Hooke (b) Lamark
   (c) De vries (d) None of these.

121. Name the largest tree of the world
(a) Palm (b) Ashoka
(c) Sequoia Gigantica (d) None of these.

122. Which is the edible part of the cauliflower
(a) Petal (b) Pistel
(c) Stem (d) Inflorescence.

123. Name the family to which mango belongs
(a) Anacardiaceae (b) Linus
(c) Robert Hooke (d) None of these.

124. Name the scientist who is considered as the father of taxonomy
(a) Linus Panling (b) Carolus Linnaeus
(c) Robbert openheimer (d) J.C. bose.

125. With what is Syviculture concerned ?
(a) Agriculture (b) Monoculture
(c) Timber (d) None of these.

126. Name the edible part of Jack fruit
(a) Stem (b) Root
(c) Bark (d) Perianth.

127. By whom was the cohesion tension theory proposed?
(a) Dixon (b) Joly
(c) Both these (d) Cohen.

128. What does the "Corolla" in a flower comprise of ?
(a) Leaves (b) Floral
(c) Petals (d) None of these.

129. Name the largest plant in the world.
(a) Sequoia Gigentica (b) Eucalyptus
(c) Palm (d) Ashoka.

130. Name the vegetable whose botanical name is Brassica aleracea.
(a) Onion (b) Tomato
(c) Potato (d) Cabbage.

131. Name the male and female reproductive organs of a flower.
(a) Stamen (b) Pistil
(c) Either A and B (d) Both A and B

132. Arbariculture is the cultivation of .
(a) Rice (b) Trees
(c) Fruits (d) Flowers.

133. Propogation of hybrid roses is due to
(a) Grafting (b) Budding
(c) Layering (d) Air layering.

134. The edible part of apple fruit is called
(a) Mesocarp (b) Pendunc le
(c) Thallus (d) Endocarp.

135. Which one of the following is rich in ascorbic acid
(a) Orange (b) Apple
(c) Banana (d) Mango.

136. Which of the following grows under trees ?
(a) Caster (b) Cabbage
(c) Peanuts (d) Bengal gram.

137. Which one of the following is a green manure crop?
(a) Linseed (b) Barseam
(c) Mustard (d) Castor.

138. Which of the following is not a bio-fertiliser
(a) Guano (b) Green manure
(c) Urea (d) All of the above.

139. Which of the following adds nitrogen to the soil?
(a) Gram (b) Wheat
(c) Rice (d) Cotton.

140. Which of the following is not a source of energy ?
(a) Coal (b) Forest
(c) Water (d) Soil.

141. Environment pollution that can cause birth defeats is
    (a) Sulphur dioxide (b) Radioactivity
    (c) Tetrogen (d) Carbon monoxide.

142. Maize protein is deficient in the amino acid
    (a) Lysin (b) Methinine
    (c) Leucine (d) Histioline.

143. The percentage composition of carbon dioxide in the expired air is about
    (a) 2 (b) 3
    (c) 4 (d) 5.

144. A cross of hybrid with one of its parents is termed as
    (a) Biotype (b) Test cross
    (c) Hybridization (d) Back cross.

145. The first artificial plant hybrid was made around 1717 by
    (a) Thomas Fairchild (b) Conteur
    (c) Borlaug (d) Thomus Flairchild.

146. Each flower consists its female part
    (a) Ovary style and stigma (b) Ovary
    (c) Stigma (d) Thalamus.

147. In plants growth may be measured by the
    (a) Thermometer (b) Meter
    (c) Auxanometer (d) Transmitter.

148. The region of growth an elongation in the root is
    (a) At the tip
    (b) A little distance away from the tip
    (c) In the middle
    (d) Its entire extent.

149. Development of organism from a gamete without fertilization is
    (a) Apocarpy (b) Parthenocarpy
    (c) Apomixis (d) Parthenogenesis.

150. In plants the growth hormone is
(a) Leuminine (b) T.S.H.
(c) Apomixis (d) Apocarpy.

151. Which is the following tissues can be used to demonstrate
(a) Stem (b) Bark
(c) Root tip (d) Antherlobes.

152. The concept of protoplasm was first introduced in botany by
(a) Hugovon Mohl (b) Koch
(c) Khurana (d) Kornberg.

153. Karyogamy means
(a) Nuclear fission (b) Cell division
(c) Nuclear division (d) Cytoplasm fusion.

154. Karyokinesis means
(a) Nuclear fission (b) Cell fusion
(c) Nuclear fission (d) Cell division.

155. Single celled plants are called
(a) Protozoa (b) Procaryotes
(c) Protista (d) Protophyta.

156. Plant cell differs from animal cell because
(a) In plant, cell chloroplast is present whereas in animal cell chlorophyll is absent
(b) In plant cell, cell wall is present whereas it absent in animal cell
(c) In plant cell, wall is absent whereas in animal cell wall is absent
(d) None of the above.

157. Of the following which has maximum vitamin "C" ?
(a) Orange (b) Mango
(c) Apple (d) Milk.

158. Who invented cerscograph?
(a) Edison (b) Faraday
(c) Rutherford (d) J.C. Bose.

159. Where is pollen manufactured in the flower?
(a) Androecium (b) Anther
(c) Gynoecium (d) None of the above.

160. Plants step taking carbon dioxide during which time
(a) Dawn (b) Day-time
(c) Night time (d) Twilight.

161. Which of the following crops can be grown on tropical marshey as well as low lying areas
(a) Rice (b) Potato/Oat
(c) Wheat (d) Barley.

162. Which of the following contains highest protein content per gram
(a) Ground nut (b) Wheat
(c) Apple (d) Soyabean.

163. Which crop has longest interval between sowing and repairing it
(a) Wheat (b) Maize/Rice
(c) Sugarcane (d) Barley.

164. Which of the following is not a cereal ?
(a) Rice (b) Barley
(c) Oilseeds (d) Wheat.

165. Textiles, fibre is extracted from which part of cotton plant
(a) Leaves (b) Cotton seeds
(c) Root (d) Stem.

166. The green colour of leaves is due to the presence of
(a) Metallic ions contained in the soil
(b) Haemoglobin
(c) Chlorophyll
(d) All of the above.

167. The growth of seeding plans of ten transplantation will not always be proper because
(a) Roots cannot penetrate deep in to the soil

(b) During transplantation root hairs get damaged
(c) New soil may not contain the required minerals
(d) Of all the factors stated above.

168. Which of the following crop helps in nitrogen fixation?
(a) Beans (b) Wheat
(c) Rice (d) Maize.

169. Reserpine derived from the plant 'Serpentine' is used
(a) Alleviate pain
(b) Alleviate low blood pressure
(c) Alleviate high blood pressure
(d) Cure rickets.

170. The region of growth on elongation in the root is
(a) At the top
(b) At its entire extent
(c) A little away from the top
(d) In the middle.

171. Each flower consists its female part in
(a) Ovary, style and stigma
(b) Stigma
(c) Ovary, style and stigma
(d) Thalamus.

172. Which is not an oilseed crop?
(a) Sunflower (b) Sorghum
(c) Mustard (d) Linseed.

173. Which is the rich source of virtamin 'C' ?
(a) Citrus fruits (b) Eggs
(c) Milk (d) Green vegetables.

174. During day time plants absorb
(a) Nitrogen
(b) Oxygen
(c) Carbon dioxide
(d) None of the above.

175. Viruses usually inject the whole plant except the

(a) Stem apex (b) Cortex
(c) Pith (d) Phloem.

176 True nucleus is absent in

(a) Green algae (b) Fungi
(c) Lichens (d) Bacteria.

177. Which of the following groups of plants are highly useful in increasing soil fertility ?

(a) Red algae (b) Fungi
(c) Bacteria (d) Bryophytes.

178. One of the following causes cholera

(a) Bacills (b) Commas
(c) Spirilia (d) Cocci.

179. Algae are in the same major groups of plants as are the

(a) Mosses (b) Liver warts
(c) Fungi (d) Ferns.

180. All fungi are

(a) Autotrophs (b) Saprophytes
(c) Heterotrophs (d) Parasites.

181. White rust of crucifers is caused by

(a) Albugo (b) Phytopthora
(c) Puccinia (d) Ustelago.

182. Lichens are brown as pioneers of vegetation in a

(a) Hydrosere (b) Halosere
(c) Psaimmosere (d) Lithosere.

183. Protonema is a stage in the life cycle of

(a) Riccia (b) Funaria
(c) All brophytes (d) Cycas.

184. Gametophytes is most developed in

(a) Moss (b) Fern
(c) Cycas (d) Pinus.

185. Rhizophore of selaginella is

(a) Root (b) Sterm
(c) New organ (d) None of the above.

186. The number of spores in the sparanglium of fern is

(a) 16 (b) 32
(c) 64 (d) 128.

187. Which of the following is a living fossil?

(a) Wind (b) Water
(c) Insect (d) man.

188. Perisperm is

(a) Surviving nucellus in the seed
(b) Out growth of the funiculus
(c) Outgrowth of the stem
(d) Out growth of the peristone.

189. The branch of botany which deals with the form of the plant is known as

(a) Physiology (b) Morphology
(c) Ecology (d) Taxonomy.

190. Red root is another name of

(a) Sugarcane roots (b) Bat root
(c) Potato roots (d) None of the above.

191. Cuscuta is a

(a) Total stem parasite
(b) Partial stem parasite
(c) Total root parasite
(d) Partial root parasite.

192. Seeds develop from

(a) Ovules (b) Ovaries
(c) Anthers (d) Pistils.

193. Synergids are

(a) Haploid (b) Diploid
(c) Triploids (d) Tetraploid.

194. The bacterial name of Bajra is
(a) Zea mays (b) Peninsetum typhoides
(c) Sarghum vulgare (d) Triticum satvum.

195. India ranks first in production of ?
(a) Sweet potato (b) Tea
(c) Corn (d) Potato.

196. The commercial jute fibres are
(a) Phloem fibres (b) Xylem fibres
(c) Intexlary fibres (d) None of these.

197. Tapioca belongs to
(a) Rubiacea (b) Apocyanacea
(c) Euphorbiacea (d) Chenopodiacea.

198. The word "protoderm" was coined by.
(a) Hanstein (b) Haberlandt
(c) Schmidt (d) Nageli.

199. Companion cells are absent in
(a) Dicot stem (b) Monocot stem
(c) Dicot root (d) Pinus stem.

200. Physical basis of life is
(a) Nucleus (b) Cell
(c) Protoplasm (d) Food.

201. The difference between primary and secondary phloem is
(a) Paranchyma and fibres are present in primary phloem
(b) Parenchymal cells are organised into ray systems in secondary phloem
(c) Companion cells are absent in secondary phloem
(d) Both are a like.

202. Xylem is said to be endrach when
(a) The protoxylem is located towards the periphery
(b) The protoxylem is located towards the centre
(c) The protoxylem is located in the centre of the xylem
(d) The protoxylem gets obliterated.

203. The non-superized phloem like cells produced by phellogen called

(a) Powdery cells (b) Complimentary cells
(c) Supplementary cells (d) Parenchynonal cells.

204. Active absorption of ions is facilitated by

(a) Oxygen
(b) A.T.P.
(c) More effective plasma membrane
(d) Slightly higher temperature.

205. Doramancy in buds and seeds is due to

(a) Indole acetic acid (b) Gibberellin
(c) Cytokinin (d) Abscisic acid.

206. Rhizomes are

(a) positively geotropic (b) plagiotropic
(c) diageotropic (d) negatively geotropic.

207. Carotenoids

(a) Do not take any part in photosynthesis
(b) Also absorb light energy and transfer it to chlorophyll a
(c) Absorb only heat energy
(d) Help to colour the young leaves and pericarps.

208. Number of A.T.P molecules produced during anaerobic glycolysis is

(a) One (b) Two
(c) Three (d) Four.

209. In the list give below who considered as the "Father of Taxonomy"?

(a) Aristotle (b) Linnaeus
(c) John Ray (d) Hippocrates.

210. Hutchinsons classification of plants is

(a) An artificial system (b) A natural system
(c) A phylogenetic system (d) Few systematic.

211. Ca·age is
(a) The entire plant (b) The large axillary bud
(c) The large terminal bud (d) The midified shoot.

212. Which of the following is the correct pair ?
(a) Lathyrus -Caesalpineae
(b) Asparagus- Compositeae
(c) Raptaranus- Cruciferance
(d) Tomato-Papilionaceae.

213. Gobar gas contains mainly
(a) Ethylene (b) Methane
(c) Hydrogen and ethylene (d) Hydrogen and oxygen.

214. Opuntia was menance a few decads back in india it was almost eradicated by
(a) Sparaying DDT
(b) Using tractors
(c) Cochineal insect introduction
(d) Using 2-4 D.

215. Deplection cause forests results in
(a) Less rainfall (b) Erosion
(c) Loss of fertility (d) All the above.

216. Automobiles cause air pollution. Fumes caontain
(a) Carbon dioxide (b) Carbon monoxide
(c) Sulpher dioxide (d) Water vapour.

217. The most widely existing pollutant in the air is
(a) Carbon dioxide (b) Carbon monoxide
(c) Sulphur dioxide (d) None of these.

218. Golgi bodies help in
(a) Respiration
(b) Initaiting cell division
(c) Synthesising secretary substances
(d) Protein synthesis.

219. The nucleus has
(a) One membrane with pores
(b) Two membranes with pores
(c) Two membranes with pores through which substances do not pass
(d) Two membranes with pores through.

220. Which macromolecules may pass Drosophila the fruit fly is
(a) Found only in USA.
(b) Not seen in India
(c) Found all over the world
(d) Confined only to temperate regions.

221. Skin colour is determined by cumulative genes hypothesis is designed by
(a) Bateson (b) Benzor
(c) Davenport (d) Hicks.

222. Beet is
(a) A root (b) A stem
(c) A stem and root (d) An entire plant.

223. Rice is
(a) A terrestrial plant (b) A inter chemicals
(c) A semi-aquatic plant (d) An amphibious plant.

224. The longest wheat producing state in India is
(a) Punjab (b) U.P.
(c) Haryana (d) M.P..

225. Which state ranks first in the production of Rice
(a) Tamil Nadu
(b) Andhra Pradesh
(c) Madhya Pradesh
(d) West Bengal.

226. The immatured stage eggs are called as
(a) Microlececithal (b) Oogenisis
(c) Oocyte (d) Zygote.

227. A specialized product of an organism adopted to take part in fertilization is
(a) Gamete (b) Zygote
(c) Mitochondria (d) R.N.A..

228. Mark the correct point. Meristenatic cells are characterised by
(a) Thick walls
(b) Large nucleus
(c) Presence of many mitochondria
(d) Large vacuoles.

229. In grasses stem increases in length by
(a) Apical meristem (b) Secondary meristem
(c) Intercalary meristem (d) Lateral meristem.

230. The main difference between the living and non-living is
(a) In the growth
(b) In the size
(c) In the movement
(d) In the presence of protoplasm.

231. Physical basis of life is
(a) Protoplasm (b) Food
(c) Cell (d) Nucleus.

232. Colloidal theory for protoplasm has been given by
(a) Altmann (b) Flemming
(c) Eicher (d) Butschli.

233. Protoplasm is a
(a) Calloidal solution
(b) Suspension
(c) True solution
(d) Polyphasic calloidal system.

234. National Institute of Oceanography is situated at
(a) Mumbai (b) Panji (Goa)
(c) Coimbatore (d) Chennai.

235. Royal Botanical Garden is situated in
(a) Dehradun (b) Darjeeling
(c) Kew (d) New York.

236. largest undevided leaf is found in
(a) Musa sapientum (b) Phoenix elactylifera
(c) Alocacia macrorrhiza (d) Cythea (treefera).

237. Largest tree of the world is
(a) Ficus (b) Pinus
(c) Sequoia (d) Eucalyptus.

238. The plant species which grow at the slowest rate is
(a) Pinus sylvestries (b) Oppuntia spp.
(c) Ficus religiose (d) Dioon edula.

239. Christmas tree is
(a) Treefern (b) Douglas fir
(c) Dwarf pinus tree (d) Cyeas revolution.

240. Oldest living plant of the world is
(a) Pinus aristola (b) Sequoia
(c) Indian banyan (d) Australian eucalyptus.

241. Cell Organells are embeded in
(a) Cytoplasmic (b) Mucein
(c) Protoplasm (d) Cytoplasm.

242. Which of the following proteins are non-structural and non-fibrillar in nature?
(a) Myosin (b) Keratin
(c) Collagen (d) Globulin.

243. Which of the following properties of a calloid are shared by the protoplasm ?
(a) Contractibility (b) Reverseibility
(c) Elasticity (d) All of the above.

244. Which element has the highest percentage in cytoplasm?
(a) Carbon (b) Hydrogen
(c) Oxygen (d) Nitrogen.

245. The compound present in largest quantity in an inttactive protoplasm is-

(a) Glucose (b) Fat
(c) Protein (d) Water.

246. The substance which make up about 30% of cytoplasm in plant cells is-

(a) Minerals (b) Water
(c) Protein (d) Fats.

247. Which of the following is the chief organic substance present in cytoplasm

(a) Protein (b) Sugar
(c) Water (d) Liquid.

248. Who defined protoplasm as a physical basis of life ?

(a) J. Hucley (b) Watson
(c) Dujardian (d) Schwann.

249. The Conversion of plasmosal to plasmagel is a

(a) Physical phenomenon
(b) Chemical phenomenon
(c) Physico chemical phenomenon
(d) Lichona.

250. The branch which deals with study of cell structure is known as

(a) Histology (b) Ecology
(c) Morphology (d) Cytology.

251. True nucleus is absent in

(a) Ribosomes (b) Green algae
(c) Fungi (d) Lichens.

252. A plant cell usualy differs from an animal cell in the absence of

(a) Ribosomes
(b) Centriole
(c) Mitochondria
(d) Endoplasmic reticulum.

253. Prokaryotic cell does not possess
(a) Cytoplasm (b) Plasma membrane
(c) Nuclear membrane (d) Cell wall.

254. Cell theory was propounded by-.
(a) Schliden and schwann (b) Robert hooke
(c) Robert Koch (d) Darwin.

255. Which of the following is an exception to cell theory?
(a) Fungi (b) Bacteria
(c) Virus (d) PPLO or mycoplasma.

256. What is the best estimate of average with of an onoin cell?
(a) 0.007 mm (b) 0.023 mm
(c) 0.075 mm (d) 0.100 mm.

257. The first living cell was possibly like
(a) Monera (b) Protista
(c) Bacteria (d) Virus.

258. Which of the following is the smallest plant cell?
(a) PLO (b) Parenencyma
(c) Bacteria (d) Virus.

259. "Cell is a unit of life" was proposed by
(a) Malpighi (b) Leuwenhook
(c) Schleiden (d) Robert Hocke.

260. Who has applied cell theory to plant ?
(a) Jonsen (b) Schleiden
(c) Schwann (d) Swanson.

261. In a plant cell vacuole contains
(a) Water & dissolved substances
(b) Only air
(c) Water
(d) Cytoplasm.

262. The term cell was coined by
(a) Robert Hocke (b) Robert Brown
(c) Flemming (d) Leuwenhock.

263. Plasmalamma is another term for
(a) Microfibils (b) Cell wall
(c) Plasmamembrane (d) Middlelamella.

264. Cell wall is
(a) Permeable
(b) Impermeable.
(c) Semi-permeable
(d) Selectively permeable

265. The outer most portion of a plant cell is
(a) Ectoplast (b) Cell wall
(c) Cuticle (d) Cell membrane.

266. The inner most layer of the wall is
(a) Middle lamella (b) Plasma membrane
(c) Tertiary wall (d) Secondary wall.

267. The peptidoglycan polymer forms the wall
(a) Fungi (b) Eukaryotes
(c) Prokaryotes (d) Higher plants.

268. Lignin is the main constituents of
(a) Primary wall
(b) Secondary wall
(c) Middle wall
(d) Plasmalemma/Pit membrane.

269. How the fibrils cellulose are cemented together to form elastic cell wall? They are cemented with the help of
(a) Chitin (b) Pectine
(c) Lignin (d) Suberin.

270. Pectine is a polymer of
(a) Glucose (b) Mannose
(c) Pactinic alginic acid (d) Galacturonic acid.

271. The torus of a bodred pit is made up of
(a) Phospholipids (b) Glucose
(c) Callulose and cutin (d) Suberin.

272. Which type of pit is characterised by the presence of torus?
(a) Simple pit (b) Bardered pit
(c) Blind pit (d) All of these.

273. Sudan IV is used to stain
(a) Cutin (b) Cellulose
(c) Suberin (d) Lignin.

274. Seed coat of date palm is made up of
(a) Cellulose (b) Hemicellulose
(c) Lignin (d) Pactin.

275. The chief component of middle lamella in plant cell is
(a) Mg (b) Ca
(c) Na (d) K.

276. The chemical substance most abundantly present in the middle lamella is
(a) Pectin (b) Lignin
(c) Subrin (d) Cutin.

277. An example of a cell lacking nucleer membrane and mitochondria is
(a) Protozoan cell (b) Bacteria cell
(c) Algae cells (d) Fungal cells.

278. The cell wall of plants is made up of cellulose. This is believed to be a
(a) Amino acid (b) Polysaccharide
(c) Liquid (d) Protein.

279. The hydrophobic ends of phospholipid molecules are
(a) Polar (b) Non polar
(c) Bipolar (d) Neutral.

280. Phospholipid molecules are
(a) Alkaline (b) Acidic
(c) Neutral (d) Amphipathic.

281. Longest dissected leaves are found in
(a) Date palm (b) Raffia palm
(c) Fern palm (d) Tree fern.

282. How many regions are in India under MAB programme in tiger project?

(a) 09 (b) 13
(c) 17 (d) 51.

283. The father of Indian Botany is

(a) Birbal Sahni (b) E.J. Butler
(c) William Rexburgh (d) Father H. Santapau.

284. What is the botanical name of radish

(a) Raphanus sativus (b) Hibiscus rosasinesis
(c) Mangi ferrendica (d) None of these.

285. Who is father of Mycology ?

(a) Alexender Flemming (b) De Bary
(c) K.C. Mehta (d) Fritsch.

286. Who is father of Broyalogy?

(a) Linnaeur (b) Robbert Brown
(c) Johnson Hedwing (d) De Candalle.

287. Word 'taxonomy' was coined by

(a) Thephrastur (b) A. P. de Candalle
(c) Randle (d) Linnaeur.

288. M.O.P Iyenger is known for his contributions in

(a) Mycology (b) Phycology
(c) Oncology (d) Bacterialogy.

289. Woksman is known for his contributions in

(a) Mycology (b) Phycology
(c) Bacteriology (d) Virology.

290. Ringworm of the scalp is caused by

(a) Microsporum (b) Penicillium
(c) Mucor (d) Albugo.

291. The presence of dalipore septum is a characteristic feature of

(a) Phycomycetes (b) Arcomyctes
(c) Basidiomycetes (d) Myxomycetes.

292. Wart disease of patate in caused by
   (a) Pythium deparyanum
   (b) Phytophothora infertans
   (c) Paronaspora destructor
   (d) Synehytrium indobioticum.

293. Downy mildew disease is caused by
   (a) Peronospora sp.
   (b) Puccinia
   (c) Albugo sp.
   (d) Phytaphthora sp.

294. Powedery mildew crop is caused by
   (a) Ascomycetes (b) Basidiomycetes
   (c) Phycomycetes (d) Bacteria.

295. Which of the following cause must disease in cutivated plants ?
   (a) Puccinia (b) Cystopus
   (c) Ustilago (d) Agaricas.

296. Which of the following fungus is poisonous ?
   (a) Albugo (b) Erysiphe
   (c) Amanita (d) Saccharomyces.

297. In which of the following aflatoxin was first discovered?
   (a) Paccunia (b) Aspergillus
   (c) Mucor (d) Albugo.

298. Aflatoxin is follwoed by
   (a) A virus (b) A fungus
   (c) A bacterium (d) A nematode.

299. Heterothallism in fungi was discovered by
   (a) Darwin (b) Bessey
   (c) Hocke (d) Blackeslee.

300. In India paper is mainly manufactured from
   (a) Soft wood conifers (b) Bamboos
   (c) Eucalyptus (d) Waste paper.

301. The first geneticist was
   (a) Engler (b) Mendel
   (c) Schwann (d) Miller.

302. Which of the following is known as the father of Genetics ?
   (a) C. Darwin (b) Weismann
   (c) Mendel (d) Hugo de Vries.

303. The first scientific studies in Genetics were carried out by
   (a) Mendel (b) Darwin
   (c) Hugo de vries (d) Lamarck.

304. Mendel was a
   (a) Physiologist (b) Plant breeder.
   (c) Cytologist (d) Taxonomist

305. The first scientific study leading to formulation of laws of inheritance was carried out by
   (a) Darwin (b) Hugo de vries
   (c) Lamarck (d) Mendel.

306. The name of the discipline dealing with the study of inheritance is
   (a) Cytology (b) Evolution
   (c) Genetics (d) Anatomy.

307. Genes are located in a linear fashion on the
   (a) Nuclear membrane (b) Nucleolus
   (c) Plasma membrane (d) Chromosomes.

308. Which of the following is the unit of inheritance ?
   (a) Phenotype (b) Genotype
   (c) Gene (d) None of these.

309. The term gene was given by
   (a) T.H. Margon (b) Mendel
   (c) W.L. Jansen (d) Hugo de vries.

310. Term genetics and allele was coined by
   (a) Morgan (b) William Bateson
   (c) Wilhelm Johanssen (d) Carl Carrens.

311. Who established that gene was a discrete unit of the chromosomes?
(a) Mendel (b) Morgan
(c) Hugo de vries (d) Batesoh.

312. Genes do not occur in pairs in
(a) Root (b) Embryo
(c) Spore (d) Nucellus.

313. Weismann propounded-
(a) Vital force theory (b) Gene pool theory
(c) Germ plasm theory (d) Protoplasm theory.

314. Who proposed the hypothesis that every organ of an animal body emits some sort of vapours which unite to form a new individual?
(a) Lamarck
(b) Pythagoras
(c) Leonardo da vinai
(d) William Bateson.

315. Who suggested that sexual cells are miniature copies of the adults called mankins ?
(a) Kobreuter (b) Swammerdam
(c) Weisman (d) Maupertius.

316. "Like begest like" is an important and universal phenomenon of life. This is due to
(a) Dominance (b) Genetics
(c) Inheritance (d) Fugencies.

317. Mendel conducted his hybridization experiment with
(a) Wild pea (b) Chick pea
(c) Pigeon pea (d) Garden pea.

318. An exception to Mendel's law is
(a) Independent assortment
(b) Dominance
(c) Purity of genetes
(d) Linkage.

319. Mendel choose pea plants because
   (a) They were cheap
   (b) They were having contrasting characters
   (c) They were easily available
   (d) Of their great economic importance.

320. Mendel worked in
   (a) 17 th century (b) 18 th century
   (c) 19 th century (d) 20 century.

321. Mendel is famous for his work as
   (a) Pisum (b) Drosophila
   (c) Neurospara (d) Ochothera.

322. Allelomorphs are
   (a) Pairs of contrasting characters
   (b) Pairs of determiners
   (c) Segregation of characters in pairs
   (d) None of these.

323. By definition as allele is
   (a) The alternative state of two genes
   (b) The alternative state of one gene
   (c) Different genes representing different characters
   (d) Different genes representing one character.

324. Mendel did his experiment on
   (a) Sweet pea
   (b) Garden pea
   (c) Pigeon pea
   (d) Phaseolus radiations.

325. On which plant Mendel had carried out his investigations?
   (a) Sweet pea (b) Wild pea
   (c) Cow pea (d) Pigeon pea.

326. Who gave the laws of inheritance?
   (a) Morgan (b) Dalton
   (c) Mendel (d) Shull.

327. Gregor Mendel propunded ?
   (a) Theory of mutation
   (b) Theory of the inheritance of aquired characters.
   (c) Law of heredity
   (d) Cell theory in plants.

328. Mendel was born in
   (a) Austria (b) Brunn
   (c) Silisian village (d) Vienna.

329. Who is father of heredity and genetics or the great genetics was
   (a) Khurana (b) Josph Kalreuter
   (c) Gregor Mendel (d) Carl Carrens.

330. Plants can be made disease resistant by
   (a) Colchicine treatment
   (b) Breeding with their wild relatives
   (c) Hormone treatment
   (d) X-rays treatment.

331. Gregar Mendel is famous for propounding
   (a) The cell theory
   (b) The laws of heredity
   (c) The theory of mutation
   (d) The theory of inheritance of acquired characters.

332. Mendel conducted hybridization experiment with
   (a) Chick pea (b) Pigeon pea
   (c) Garden pea (d) Wild pea.

333. In 1900 A.D. three biologists independent Mendel's principles. They are
   (a) De vries, Carrens and Tschermark
   (b) Sutton, Morgan and Bridges
   (c) Avery, Maclead and Mc Carty
   (d) Batson, Puneet and Bridges.

334. Mendels laws aply when
  (a) F_1 in monohybrid gross shows two types of individuals
  (b) The characters are linked
  (c) Parents are pure breeding
  (d) First pair of constrating characters dependent upon other pairs.

335. Which of the following was proposed by Mendel?
  (a) Incomplete dominance (b) Dominance
  (c) Segregation (d) Independent assortment.

336. The individuals of one maternal and paternal parents are known as
  (a) Epistats (b) Hypostats
  (c) Siblings (d) Gynander.

337. Emasculation is achieved by
  (a) Removal of anthers (b) Removal of style
  (c) Removal of petals (d) Removal of sepals.

338. Who presented Mendelian findings as law?
  (a) John Gregar Mendel (b) Morgan
  (c) Johanssen (d) Carl Correns.

339. Who and which animal the Mendellian laws of inheritance were first confirmed?
  (a) Carrens on dogs (b) Morgan on drosophila
  (c) Sessyneg on albino rats (d) Muller on guinea pigs.

340. Who rediscovered Mendel,s work and brought it to light ?
  (a) De vries of Holand
  (b) Tschermack sessyneg of Australia
  (c) Carl Carrens of Germany
  (d) All seperately.

341. Heterosis is
  (a) Hybrid incompatibility
  (b) Hybrid vigour
  (c) Hybrid sterility
  (d) Structural hybridity.

342. A cross between allelic or unlike organism is
   (a) Test cross (b) Back cross
   (c) Hybridization (d) Heterosis.

343. Hybridization means
   (a) Removal of stemens
   (b) Production of large sized grains
   (c) Mixing parental
   (d) Study of pollination habit.

344. The character which appears at F_1 generation is
   (a) Dominant (b) Recessive
   (c) Incomplete dominant (d) incomplete recessive.

345. If an allelic pair of gene is present in sweet pea it would prove the law of
   (a) Dominance (b) Segregation
   (c) Independent (d) Linkege.

346. The character which becomes physically apparent is called
   (a) Gene (b) Genotype
   (c) Phenotype (d) Homozygous.

347. A complete set of chromosomes inherited as a unit from one parent is known as
   (a) Karyotype (b) Gene pool
   (c) Genome (d) Genotype.

348. The genetic constitution of an organism is known as
   (a) Genotype (b) Phonotype
   (c) Genome (d) Gene pool.

349. The term genotype was given by
   (a) H.J. Muller (b) W.S. Sutton
   (c) W.L. Johansen (d) T.H. Morgan.

350. If a plant heterozygous for follows is selfed the F_2 generation has both tall and dwarf plants this proves the principles of
   (a) Dominance
   (b) Segregation

(c) Independent assortment
(d) Incomplete dominance.

351. The product of hybridization is known as
(a) Clone (b) Homozygous
(c) Hybrid (d) Heterozygous.

352. Linolenic acid is unsaturated fatty acid and its content is highest in
(a) Cotton oil (b) Groundnut oil
(c) Sunflower oil (d) Coconut oil.

353. Enzymes which help in electron transfer are
(a) Cytochromes (b) Isomerases
(c) Proteases (d) All of these.

354. Desired improved variety of economically useful crops are raised by
(a) Mutation (b) Hybridization
(c) Biofertilizers (d) Natural selection.

355. Which of the following is haploid in gymnosperms ?
(a) Pollen grain, mega spore and root
(b) pollen grain, mega spore and nucleus
(c) Mega spore, mother cell, root and leaf
(d) Endosperm, pollen grain and megasporl.

356. Root nodules for nitrogen fixation in leguminous trees are produced by species of the genus
(a) Rhizobium (b) Azotobacter
(c) Frankea (d) Thiobacillus.

357. The structure of chromosome can be best seen at
(a) Prophase (b) metaphase
(c) Phase (d) telaphase.

358. Pteridophytes differ from bryophytes in possessing.
(a) Spores (b) Archegonia
(c) Vascular tissue (d) All of these.

359. The best method for the control of plant disease is
   (a) Gradication
   (b) Solar treatment
   (c) Using disease resistant varieties
   (d) Using fungicide.

360. Emasculation is concerned with
   (a) Hybridization (b) Clonal selection
   (c) Mass selection (d) Pure line selection.

361. Photophosphory lation was discovered by
   (a) Hill (b) Arnds
   (c) Calvin (d) Ruben and Kamen.

362. Pure line breed refers to
   (a) Homozygosity only
   (b) Heterozygosity and linkage
   (c) Heterzygosity only
   (d) Homozygosity independent assortment.

363. Pneumatophores organs are used in
   (a) Respiration (b) Photosynthesis
   (c) Protein synthesis (d) Secretion of salt.

364. When vegetative cells of zygote form embryo, it is called
   (a) Apospary
   (b) Apomixis
   (c) Diploid polyembryony
   (d) Adventive polyembryony.

365. Parasite alga is
   (a) Volvox (b) Vaucheria
   (c) Cephaleuros (d) Oedogonium.

366. The phenomenon in which an allele of an gene suppresses the activity of an allele of another gene is known as
   (a) Suppression (b) Inactivation
   (c) Epistasis (d) Dominance.

367. Law of independent assortment was proposed by

(a) Mendel (b) Linnacus
(c) Robert Hook (d) Melvin Kelvin.

368. The growth harmones responsible for 'olting' are

(a) Auxins (b) Kinetins
(c) Coumarins (d) Gibberellins.

369. The factor which is not limiting in normal conditions for photosynthesis is

(a) Light
(b) Carbon dioxide
(c) Water in plant cell
(d) Chlorophyll content of leaf.

370. An alga very rich in protein is

(a) Spirogyra (b) Ulothrix
(c) Chlorella (d) Nostoc.

371. The function of a vessel is

(a) Conduction of food only
(b) Conduction of water and minerals
(c) To provide mechanical strength
(d) All of these.

372. Atmospheric nitrogen fixation is carried on by

(a) Funaria (b) Anabaena
(c) Chlamydomonas (d) Fern gametophyte.

373. Protein synthesis takes place in

(a) Mitochondria (b) Ribosomes
(c) Chloroplast (d) Golgi bodies.

374. Gills are present in

(a) Agaricus (b) Spirogyora
(c) Moss (d) Fern.

375. Study of farm and structure of an organism is called

(a) Anatomy (b) Taxonomy
(c) Ecology (d) Marphology.

ANSWER SHEET

1.	(a)	2.	(c)	3.	(b)	4.	(c)
5.	(d)	6.	(d)	7.	(a)	8.	(a)
9.	(c)	10.	(c)	11.	(a)	12.	(b)
13.	(d)	14.	(c)	15.	(a)	16.	(c)
17.	(c)	18.	(c)	19.	(b)	20.	(c)
21.	(c)	22.	(a)	23.	(d)	24.	(c)
25.	(d)	26.	(b)	27.	(a)	28.	(c)
29.	(b)	30.	(b)	31.	(a)	32.	(a)
33.	(b)	34.	(b)	35.	(d)	36.	(b)
37.	(c)	38.	(d)	39.	(c)	40.	(b)
41.	(a)	42.	(b)	43.	(a)	44.	(c)
45.	(b)	46.	(b)	47.	(b)	48.	(b)
49.	(d)	50.	(c)	51.	(c)	52.	(c)
53.	(c)	54.	(c)	55.	(a)	56.	(a)
57.	(b)	58.	(d)	59.	(c)	60.	(b)
61.	(b)	62.	(c)	63.	(c)	64.	(c)
65.	(a)	66.	(b)	67.	(b)	68.	(a)
69.	(c)	70.	(a)	71.	(c)	72.	(a)
73.	(c)	74.	(b)	75.	(a)	76.	(a)
77.	(c)	78.	(d)	79.	(c)	80.	(d)
81.	(d)	82.	(d)	83.	(c)	84.	(c)
85.	(d)	86.	(b)	87.	(d)	88.	(d)
89.	(c)	90.	(b)	91.	(c)	92.	(b)
93.	(c)	94.	(c)	95.	(d)	96.	(c)
97.	(a)	98.	(a)	99.	(d)	100.	(d)
101.	(b)	102.	(b)	103.	(a)	104.	(b)
105.	(c)	106.	(b)	107.	(b)	108.	(b)
109.	(a)	110.	(a)	111.	(c)	112.	(a)
113.	(a)	114.	(c)	115.	(a)	116.	(b)
117.	(a)	118.	(c)	119.	(b)	120.	(c)
121.	(c)	122.	(d)	123.	(a)	124.	(b)
125.	(c)	126.	(d)	127.	(c)	128.	(c)
129.	(a)	130.	(d)	131.	(d)	132.	(b)
133.	(a)	134.	(a)	135.	(b)	136.	(c)
137.	(b)	138.	(d)	139.	(a)	140.	(a)
141.	(b)	142.	(b)	143.	(d)	144.	(d)
145.	(a)	146.	(a)	147.	(c)	148.	(a)
149.	(d)	150.	(c)	151.	(c)	152.	(a)
153.	(a)	154.	(d)	155.	(c)	156.	(b)

157.	(a)	158.	(d)	159.	(b)	160.	(c)
161.	(a)	162.	(a)	163.	(d)	164.	(c)
165.	(b)	166.	(c)	167.	(b)	168.	(a)
169.	(c)	170.	(c)	171.	(a)	172.	(b)
173.	(a)	174.	(c)	175.	(a)	176.	(d)
177.	(c)	178.	(b)	179.	(c)	180.	(c)
181.	(a)	182.	(d)	183.	(b)	184.	(a)
185.	(b)	186.	(c)	187.	(c)	188.	(a)
189.	(b)	190.	(b)	191.	(b)	192.	(a)
193.	(a)	194.	(a)	195.	(b)	196.	(d)
197.	(a)	198.	(c)	199.	(b)	200.	(d)
201.	(b)	202.	(b)	203.	(b)	204.	(b)
205.	(d)	206.	(c)	207.	(b)	208.	(b)
209.	(b)	210.	(c)	211.	(c)	212.	(c)
213.	(b)	214.	(c)	215.	(d)	216.	(b)
217.	(b)	218.	(c)	219.	(d)	220.	(c)
221.	(a)	222.	(c)	223.	(b)	224.	(d)
225.	(c)	226.	(a)	227.	(b)	228.	(c)
229.	(a)	230.	(d)	231.	(b)	232.	(c)
233.	(a)	234.	(b)	235.	(c)	236.	(c)
237.	(c)	238.	(d)	239.	(b)	240.	(a)
241.	(d)	242.	(d)	243.	(c)	244.	(d)
245.	(b)	246.	(a)	247.	(a)	248.	(c)
249.	(c)	250.	(d)	251.	(a)	252.	(b)
253.	(c)	254.	(a)	255.	(c)	256.	(b)
257.	(a)	258.	(a)	259.	(d)	260.	(b)
261.	(a)	262.	(a)	263.	(c)	264.	(a)
265.	(b)	266.	(c)	267.	(c)	268.	(b)
269.	(b)	270.	(d)	271.	(d)	272.	(b)
273.	(c)	274.	(b)	275.	(c)	276.	(a)
277.	(b)	278.	(b)	279.	(b)	280.	(d)
281.	(a)	282.	(b)	283.	(c)	284.	(b)
285.	(c)	286.	(b)	287.	(b)	288.	(e)
289.	(c)	290.	(a)	291.	(c)	292.	(d)
293.	(a)	294.	(a)	295.	(c)	296.	(c)
297.	(b)	298.	(b)	299.	(d)	300.	(b)
301.	(b)	302.	(c)	303.	(a)	304.	(b)
305.	(d)	306.	(c)	307.	(d)	308.	(c)
309.	(c)	310.	(b)	311.	(b)	312.	(c)
313.	(c)	314.	(b)	315.	(b)	316.	(c)
317.	(d)	318.	(d)	319.	(b)	320.	(c)

321. (a)	322. (a)	323. (b)	324. (a)
325. (a)	326. (c)	327. (c)	328. (c)
329. (c)	330. (b)	331. (b)	332. (c)
333. (a)	334. (c)	335. (a)	336. (c)
337. (a)	338. (d)	339. (b)	340. (d)
341. (b)	342. (c)	343. (b)	344. (a)
345. (a)	346. (c)	347. (c)	348. (a)
349. (c)	350. (b)	351. (d)	352. (c)
353. (a)	354. (b)	355. (d)	356. (a)
357. (b)	358. (c)	359. (c)	360. (a)
361. (b)	362. (d)	363. (a)	364. (d)
365. (c)	366. (c)	367. (a)	368. (d)
369. (d)	370. (c)	371. (b)	372. (b)
373. (b)	374. (a)	375. (d)	

4
GENERAL CHEMISTRY

1. Isotopes are atom that have
   (a) Same atomic number but different atomic mass
   (b) Same atomic mass but different atomic number
   (c) Same number of protons but different number of electrons
   (d) Same number of neutrons but different number of protons.

2. The number of electrons present in a Carbon atom is
   (a) 4 (b) 5
   (c) 6 (d) 8.

3. All bases contain the ions of
   (a) Hydrogen (b) Hydrooxide
   (c) Nitrogen (d) Chloride.

4. The percentage of water covering earth surface in the world is approximately
   (a) 40% (b) 50%
   (c) 60% (d) 70%.

5. Proton is
   (a) Positive charged (b) Negative charged
   (c) Not charged (d) None of these.

6. Carbon dioxide is helpful in burning
   (a) True (b) False
   (c) Sometime true (d) None of these.

7. Nitrogen is
   (a) Heavier than air (b) Lighter than air
   (c) Equal to the air (d) None of these.
8. A chemical name of caustic soda is
   (a) Sodium bicarbonate (b) Sodium hydroxide
   (c) Sodium nitrate (d) None of these.
9. The main ore of iron is
   (a) Haematite (b) Sinebar
   (c) Galena (d) None of these.
10. Which of the following gases is present in the air in the largest proportion?
    (a) Hydrogen (b) Nitrogen
    (c) Oxygen (d) Carbon dioxide.
11. A atom is made up of
    (a) Neutrons and Protons
    (b) Neutrons and Electrons
    (c) Electrons and Protons
    (d) Electrons, Neutrons and protons.
12. Gobar gas contains
    (a) Carbon (b) Methane
    (c) Ethylene (d) Carbon monoxide.
13. The commonest substance responsible for pollution is
    (a) Smoke (b) Carbon dioxide
    (c) Sulfer dioxide (d) Carbon monoxide.
14. If a large number of people are enclosed in a room then
    (a) Oxygen decreases and carbon dioxide increases
    (b) Oxygen increases and carbon dioxide decreases
    (c) Both oxygen and carbon dioxide decreases
    (d) Both oxygen and carbon dioxide increases.
15. Permanent hardness of water can not be removed by
    (a) Boiling (b) Adding caustic soda
    (c) Distillation (d) Adding soda.

16. The density of sea water is highest as
    (a) Depth increases and salinity increases
    (b) Depth decreases and salinity decreases
    (c) Both depth and salinity increases
    (d) Both depth and salinity decreases.

17. One letre of cold air weights heavier than the dry (1 litre) because of the
    (a) Increased number of collissions between the molicules
    (b) Increased number of molicules at low temperature
    (c) Greater energy of molecule at high temperature
    (d) Lower enrgy of molecules at high temperature.

18. Bauxite is an important ore of
    (a) Copper
    (b) Zinc
    (c) Alumminium
    (d) Iron.

19. It is advised not to sleep in a closed room where a charcol is burning because
    (a) It produce CO_2 which decreases O_2 content of atmosphere
    (b) It produces CO_2 which is injurious to health
    (c) It produces CO which is poisonous in nature
    (d) The temperature of room may rise very high.

20. Gelatin is added in ice-cream to
    (a) Hasten freezing
    (b) Give flavour
    (c) Avoid crystal formation
    (d) Act as a sweetening agent.

21. Permanent hardness of water is due to the presence of
    (a) Calcium biacarbonates
    (b) Magnesium biacarbonate
    (c) Calcium sulphate
    (d) Sodium Chloride.

22. Which of the following statement is about the metal iron is wrong?
    (a) It is the best conductor of electricity
    (b) It can be moulded to different shapes
    (c) It is magnetic in character
    (d) It is ductile.

23. Silver can be beaten in to thin plates (foil) because it is
    (a) Highly malleble (b) Highly ductile
    (c) Very costly (d) Of white colour.

24. T.N.T is a/an
    (a) Explosive (b) Drug
    (c) Fertilizer (d) Insectiside.

25. A hydrogen mollecule is made up of
    (a) 2 hydrogen (b) 3 hydrogen
    (c) Both (a) and (b) (d) None of these.

26. Which of the following is not a fossil fuel?
    (a) Charcoal (b) Neutral gas
    (c) Petroleum (d) Coal.

27. Which natural resource can be renewed by physical effects and chemical treatments?
    (a) Iron ore (b) Petroleum
    (c) Coal (d) All of the above.

28. Plaster of Paris is made from
    (a) Bauxite (b) Gypsum
    (c) Epsomite (d) Lime.

29. Raw meterial required for the manufacture of cement are
    (a) Lime, sand and soda
    (b) Lime, day, sand and soda
    (c) Lime, clay and soda
    (d) Lime and soda.

30. When clothes are bleached their colour is
    (a) Oxidised
    (b) Reduced

(c) Neither oxidised nor reduced
(d) Decomposed.

31. A physical change does not involve
(a) Change in colour (b) Change in mass
(c) Change in temperature (d) Change in pressure.

32. Which of the following represents a chemical change
(a) Melting of sugar
(b) Crystallisation of sugar
(c) Burning of sugar
(d) Dissolving sugar in water.

33. Photo chemical reaction takes place in presence of
(a) Heat (b) Light radiation
(c) Water (d) Air.

34. Reaction accompained by evolution of heat are called
(a) Endothermic (b) Exothermic
(c) Photochemical (d) Chemical.

35. Combustion of coal is a
(a) Photo chemical reaction
(b) Exothermic reaction
(c) Endothermic reaction
(d) Physical change.

36. A chemical change occurs when
(a) Sulphur melts
(b) Iron rusts
(c) Water boils
(d) Sulphur dissolves in carbon disulphide.

37. A catalyst is a substance which
(a) Increases the rate of reaction
(b) Decreases the rate of reaction
(c) Alters the rate of reaction
(d) Does not take part in the reaction.

38. Which one of the following is an element?
(a) Sodium Chloride (b) Mercury
(c) Plastic (d) Sugar.

39. Which one of the following is neither an element nor a compound ?
(a) Water (b) Air
(c) Glucose (d) Gold.

40. Which of the following is a mixture
(a) Mercury (b) Water
(c) Sea Water (d) Gold.

41. Mixtures can be seperated by
(a) Chemical method (b) Physical method
(c) By both methods (d) Chromatography.

42. Which of the following is a mixture
(a) Sugar (b) Silver
(c) Alcohol (d) Air.

43. Which of the following represents an element
(a) Oxygen (b) Ozone
(c) Marble (d) Sand.

44. Sodium of sodium chloride is a or an
(a) Mixture (b) Element
(c) Mixture (d) Element and mixture.

45. Sugar can be separated from its acquaous solution by
(a) Crystallization (b) Filtration
(c) Distillization (d) Boiling.

46. Alcohal and water may be separated by
(a) Evaporation (b) Filtration
(c) Distillation (d) Fractional distillation.

47. Isomers of an element have
(a) The same atomic number
(b) The same atomic weight

(c) The same atomic number and atomic weight
(d) Different atomic weight but same atomic number.

48. Gases react in simple ratio by
(a) Atoms (b) Atomic weights
(c) Molecules (d) Molecular weights.

49. The most abundant constituents of air is
(a) Oxygen (b) Nitrogen
(c) CO_2 (d) Moisture.

50. Nitrogen was discovered by
(a) Scheele (b) Lavoiser
(c) Rutherford (d) Cavendish.

51. Amount of N_2 in air volume is about
(a) 50% (b) 80%
(c) 20% (d) 25%.

52. Element that fails to form compound is
(a) P (b) N
(c) Sb (d) Ba.

53. The percentage of O_2 by volume in the air is
(a) 20.99 (b) 23.00
(c) 29.50 (d) 81.82.

54. The most abundant of the rare gases in the air is
(a) Neon (b) Argan
(c) Krypton (d) Helium.

55. Which of the following is not an inert gas?
(a) Argon (b) Xenon
(c) Radon (d) Flourine.

56. Hard water
(a) Give good lather with soap
(b) Does not give good lather with soap
(c) Give good lather without soap and oil
(d) Does not give good lather with cream.

57. Temporary hardness of water is due to presence of
    (a) Chloride of Ca and Mg
    (b) Nitrates of Ca and Mg
    (c) Bicarbonates of Ca and Mg
    (d) Sulphates of Ca and Mg.

58. Water has high boiling point because
    (a) Its molecules are associated by hydrogen bonds
    (b) It is mixture
    (c) It is liquid
    (d) It is mixture of O_2 and H_2.

59. Water is good solvent because
    (a) It is in liquid form
    (b) Its boiling point is very high
    (c) Its molecules are highly polar
    (d) It is dilute form.

60. Pure water is a bad conductor of electricity because it is
    (a) Only feebly ionised
    (b) A very good solvent
    (c) Highly associated by hydrogen hards
    (d) Appreciable volatile.

61. Pure water does not react with
    (a) Calcium carbonate (b) Calcium nitrite
    (c) Calcium hydride (d) Calcium Carbide.

62. The first element of the periodic table is
    (a) Sodium (b) Neon
    (c) Hydrogen (d) Helium.

63. Number of isotopes of hydrogen is
    (a) 1 (b) 2
    (c) 3 (d) 4.

64. During day time plants absorbs
    (a) Oxygen
    (b) Nitrogen.

(c) Combined mixture of O_2 and Co_2 gases
(d) Carbon dioxide.

65. Hydrogen peroxide is
(a) An oxidising agent
(b) Both oxidising and reducing agent
(c) A reducing agent
(d) None of the above is correct.

66. Nature of sodium carbonate solution is
(a) Baric (b) Acidic
(c) Neutral (d) None of the above.

67. Carbon is a
(a) Metal
(b) Non-meatal
(c) Metalloid
(d) None of the above.

68. The hardest variety of carbon is
(a) Coke (b) Gas carbon
(c) Diamond (d) Graphite.

69. Which of the following is not an all trophic form of carbon?
(a) Diamond (b) Graphite
(c) Charcoal (d) Lamb black.

70. The form of carbon known as charcol
(a) Is amarphous
(b) Consists of micro-crystals of graphite
(c) Consists of micro-crystals of graphite
(d) Is lighter than water.

71. Which of the following is not an amorphous from of carbon
(a) Coke (b) Charcoal
(c) Lamb black (d) Graphite.

72. Air becomes incombustive if its carbon dioxide content is
(a) 15 % (b) 2.5%
(c) 7.5% (d) 5%.

73. Which of the following is not a natural fertilizer?
    (a) Bone meal (b) Chile saltpertre
    (c) Urea (d) Oried blood.

74. Which of the following is not a chemical fertilizers ?
    (a) Urea (b) Ammonium nitrate
    (c) Calcium (d) Cowdung.

75. The deficiency of nitrogen in the soil
    (a) Retards plant growth and causes leaves to turn yellow
    (b) Retards plant flowering and the repening of the fruit.
    (c) Rives plant leaves a flabby dropping appearance
    (d) Prevents the seeds from germinating.

76. Which of the following ammonium compounds is not used as a nitrogenous fertilizer?
    (a) Ammonium phosphate
    (b) Ammonium sulphate
    (c) Ammonium nitrate
    (d) Ammonium phosphate.

77. Which of the following has the highest percentage of nitrogen ?
    (a) Ammonium phosphate (b) Ammonium sulphate
    (c) Calcium cynamide (d) Urea.

78. The fertilizer called super phosphate contains
    (a) $Ca\,(Po_4)_2$ (b) $CaSo_4$
    (c) $Ca\,So_4$ (d) $Ca\,(H_2Po_4)_4$.

79. Which is the complex fertilization ?
    (a) Diamnaphose $(NH_4)_2\,HPo_4$
    (b) Urea $NH_2Co\,NH_2$
    (c) Ammonium sulphate
    (d) Ammonium nitrate.

80. Which of the following crops will enrich the soil on which it is grown with nitrogen ?
    (a) Wheat (b) Suger beat
    (c) Sugercane (d) Alfalfa.

81. Which of the following statements is wrong?
   (a) Ammonium can be direct by used as nitrogenous fertilizer.
   (b) Excessive application of chemical fertilizers is not undesirables
   (c) Natural fertilizers belong to the category of complex fertilizers
   (d) Boron is needed to different extent by all plants.

82. The fertilizer called nitro-chalk is a mixture of chalk and
   (a) Ammonium nitrate (b) Sodium nitrate
   (c) Potassium nitrate (d) Nitro glycerine.

83. Plants absorb dissolved nitrates from the soil and convert them into
   (a) Free nitrogen (b) Urea
   (c) Ammonia (d) Proteins.

84. The percentage of N_2 in atmospheric air remains almost constant
   (a) Due to activity of symbiotic bacteria
   (b) Due to fixation of nitrogen
   (c) Due to nitrogen cycle in nature
   (d) Due to the effect of lightning and rains.

85. Which of the following is not an organic compound ?
   (a) Penicillin (b) Plaster of Peris
   (c) Urea (d) Oxalic acid.

86. Which of the following is not a petroleum product ?
   (a) Kerosene oil (b) Petrol
   (c) Bee wax (d) Paraffin wax.

87. Soaps are prepared from
   (a) Fats (b) Alkalis or acids
   (c) Fats & alkalis (d) Acids and fats.

88. Ordinary glass is called-
   (a) Line glass (b) Soda lime glass
   (c) Soda glass (d) None of these.

89. Ordinary glass is made by fusing together-
    (a) Sodium carbonate, silica & lime stone
    (b) Sodium shulphate & silica
    (c) Sodium nitrate and lime
    (d) Sodium nitrate and sand.

90. Cement is a mixture of -
    (a) Calcium and silicate
    (b) Tricalcium aluminate, dicalicium silicate and tricalcium silicate
    (c) Tricalcium aluminates and tricalcium silicates
    (d) Tricalcium aluminate and tricalcium.

91. Iron is made by heating iron are mixed with-
    (a) Coke and sand (b) Coke and lime stone
    (c) Coke and copper (d) Coke and scrap iron.

92. Paper is manufactured from
    (a) Wood, sodium & bleaching powder
    (b) Wood & bleaching powder
    (c) Wood calcium hydrogen sulphite & resin
    (d) Wood resin.

93. Galvanizediron is made by coating iron with
    (a) Zinc (b) Nickel
    (c) Chromium (d) Cadmium.

94. Blue-Black ink are made from
    (a) Ferrous sulphate, tannic acid & a blue dye
    (b) Ferrous sulphate & a blue dys
    (c) Ferrous sulphate & tannic acid
    (d) Ferrous sulphate & H_3So_4.

95. Atoms are mainly made up of
    (a) Protons, neutrons & electrons
    (b) Neutrons Only
    (c) Protons only
    (d) Electrons only.

96. The nucleus of an atom contains
   (a) Neutrons
   (b) Protons
   (c) Protons and Neutrons
   (d) Electrons.

97. Electrons the atomic weight of an atoms is due to
   (a) Protons
   (b) Nuetrons
   (c) Protons & neutrons
   (d) Protons, neutrons, electrons.

98. Atomic number of an element is equal to number of -
   (a) Neutrons (b) Protons
   (c) Protons & neutrons (d) Electrons & protons.

99. An atom can be defined as-
   (a) A unit negative charge
   (b) A unit positive charge
   (c) 2 units of positive charge
   (d) 2 units of negative charge.

100. A proton is identical with
   (a) An ionised hydrogen molecule
   (b) The nucleus of hydrogen atom
   (c) The nucleus of a light gas
   (d) The nucleus of a heavy gas.

101. Mass number of an element represents number of
   (a) Protons & neutrons
   (b) Protons & electrons
   (c) Electrons and neutrons
   (d) None of these.

102. Which of the following atom has no neutrons in it's nucleus?
   (a) Helium (b) Hydrogen
   (c) Lithium (d) Tritium.

103. An element with atomic number 18 is a
   (a) Noble gas (b) Alkali metal
   (c) Halogen (d) None of these.

104. Number of electrons present in the outermost orbit of a halogen is
   (a) 1 (b) 2
   (c) 3 (d) 7.

105. In an atom , electrons-
   (a) Move around the nucleus
   (b) Move around the nucleus in a circular orbit
   (c) From a diffused electron cloud around the nucleus
   (d) None of these.

106. Neutron was discoverd by-
   (a) Thomson (b) Chadwic
   (c) Rutherford (d) Bonz.

107. Energy of atomic orbitals in given shell increases in the order
   (a) Spdf (b) Sdfp
   (c) Psdf (d) Fdps.

108. The smallest part of an element which as an element is
   (a) An electron (b) A proton
   (c) A neutron (d) An atom.

109. Atomic wight of a substance is equal to-
   (a) Molecular weight of gases and volatile liquids is equal to
   (b) Equivolent weight x valency
   (c) Molecular weight x atomicity
   (d) Equivalant weight valency.

110. Molecular weight of gases and volatile liquids is equal to-
   (a) 2x vapour density (b) 1/2 x vapour density
   (c) Vapour density (d) None of these.

111. Which one of the following is the correct statement for molecular weight of carbon?
(a) 12 gm (b) 12
(c) 12mg (d) 12 moles.

112. Eequivalent weight of an element changes with-
(a) Valency
(b) Change in Cressure
(c) Change in Temperature
(d) Volume.

113. The equivalent weight is
(a) Always constant
(b) Dependent on the nature of reaction
(c) Depend upon mass
(d) All of the above.

114. The zero group elements are known as
(a) Alkali metals (b) Noble gase
(c) Alkaline eath metals (d) Transition metals.

115. The valency of the noble gas element is
(a) One (b) Two
(c) Zero (d) Three.

116. Noble gases are
(a) Monoatomic (b) Diatomic
(c) Penta atomic (d) None of these.

117. Noble gases do not occur-
(a) In atmosphere (b) In natural gases
(c) In ores (d) In sea water.

118. Most electropositive element is-
(a) Fluorine (b) Franeium
(c) Iodine (d) Potassium.

119. The most malleable metal is
(a) Mercury (b) Gold
(c) Sodium (d) Silver.

120. The most electro-negative element is
(a) Lithium (b) Fluarina
(c) Oxygen (d) Iodine.

121. The hardest substance among the following is
(a) Steel (b) Mercury
(c) Silicon carbide (d) Gold.

122. Bacing soda is
(a) K_2Co_3 (b) $NaHCo_3$
(c) Na_2Co_3 (d) Na_2So_4.

123. Which of the following metals is found free in the nature?
(a) Ba (b) Au
(c) Na (d) Ca.

124. Quick lime is
(a) $CaCo_3$ (b) CaO
(c) $CaSO_4$ $2H_2o$ (d) $CaCOH_2$.

125. Alum is -
(a) Mixed salt (b) Acidic salt
(c) Double salt (d) Amphateric salt.

126. The phenomenon of radioactivity was discovered by
(a) Bacqueral (b) Einstein
(c) Madam Curie (d) Rutherford.

127. Noble gases were discoverd by
(a) Cavendish (b) Rutherford
(c) Einstein (d) Noble.

128. B.H.C power is used as
(a) An explosive (b) A medicine
(c) An insecticide (d) A plant harmone.

129. Purest from of iron is
(a) Wrought iron (b) Steel
(c) Pig iron (d) Cast iron.

130. Steel contains-
(a) Carbon & Manganese (b) Carbon & Nickel
(c) Carbon & Iron (d) Carbon & Aluminium.

131. Common element in all the organic compounds other than H_2 is
(a) Nitrogen (b) Sulphur
(c) Phosphorus (d) Carbon.

132. Marsh gas contains large propartion of-
(a) Ethene (b) Acetylene
(c) Ethylene (d) Methane.

133. Which of the following is used as anaesthetic?
(a) Carbon tetrachloride (b) Acetone
(c) Chloroform (d) Ethyl alcohol.

134. Which of the following is used an antiseptic?
(a) Chloroform (b) Acetone
(c) Iodoform (d) Alcohol.

135. Calamine is a carbonate ore of a hard metallic element. Name the element
(a) Calcium (b) Zinc
(c) Mercury (d) Callicum.

136. What is the name give to pillars of calcium carbonate deposited on the floors of cave?
(a) Stalacites
(b) Starobites
(c) Stalagmites
(d) Skelton.

137. What % age of earth's is covered with water?
(a) 50% (b) 60%
(c) 71% (d) 90%.

138. The number of amino acids , known so far is
(a) 12 (b) 22
(c) 32 (d) 42.

139. What is the magma made of?
(a) Magnesium (b) Solid rocks
(c) Iron (d) Molten rocks.

140. Which acid is generally present in grapes?
(a) Oxalic acid (b) Citric acid
(c) Acetic acid (d) Tartaric acid.

141. Which alkaloid is found in seeds and leaves of coffee and tea plants?
(a) Codeiene (b) Caffiene
(c) Morphine (d) Leavoline.

142. Antibiotics are drugs which are used for.
(a) Bacterial diseases (b) Cancer
(c) Malaria (d) Nervous disorders.

143. What is the other name of graphite?
(a) Black lead (b) Plumbago
(c) Both (a) and (b) (d) Diamond.

144. Where is the largest deposits of platinum found?
(a) U. S.A. (b) India
(c) U.S.S.R (d) U.K..

145. By whom was the mines safety lamp invented?
(a) Davidson (b) Davy
(c) Lackart (d) Bell.

146. By whom was the telephone invented?
(a) Graham Bell (b) Edison
(c) Einstien (d) Davy.

147. Name the product obtained when heated gypsum is converted into a powder:
(a) Hydrate gypsum (b) Anhydrous gypsum
(c) Plaster of Paris (d) Barites.

148. What is brine ?
(a) Alcoholic drink
(b) Soft drink

(c) Solution of common salt
(d) Solution of Ammonia.

149. What is "sterling silver"?
(a) Alloy of silver and gold
(b) Alloy of silver and zinc
(c) Alloy of gold and zinc
(d) Alloy of silver and copper.

150. Name the country that is the largest producer of nickel
(a) India (b) Australia
(c) Canada (d) Russia.

151. By whom was dynamo invented?
(a) Faraday (b) Fleming
(c) Tess (d) Bell.

152. Which of the following has highest electron affinity?
(a) F (b) Cl
(c) P (d) I.

153. Which of the following have been prepared artificially
(a) Radom and Pelonium
(b) Uranium and Thorium
(c) Californium and Hafnium
(d) Einsteinium and Ferimium.

154. Which of the following is the strongest base?
(a) NH_3 (b) PH_3
(c) ASH_3 (d) SbH_3.

155. Which of the following solutions will haw the highest boiling point
(a) 1 % NaCl in water (b) 1 % glucose in water
(c) 1 % $CaCl_2$ (d) 1 % Sucrose in water.

156. In the reversible reaction N_2+ $3H_2$----> $2NH_3$ ------> $2NH_3$+ heat, the forward, reaction is favoured by ?
(a) Low temperature and low pressure
(b) Low temperature and high pressure

(c) High temperature and constant pressure
(d) High temperature and high pressure.

157. Which of the following is used as anti-catalyst finely divided Fe in Haber process for the manufature of NH_3

(a) CO
(b) N_2
(c) CO_2
(d) NO_2.

158. The number of atoms in a face centred cubic unit cell is?

(a) 6
(b) 4
(c) 12
(d) 8.

159. Among the following properties of alkali metals increases in magnitude as the atomic number increases?

(a) Melting point
(b) Ionic radius
(c) First lonisation energy
(d) Electronegativity.

160. Plaster of Paris in contact with water sets to a hard mass whose composition is

(a) $CaSO_4 H_2O$
(b) $CaSO_4$
(c) $CaSO_4 Ca (OH)_2$
(d) $CaSO_4\ H_2O$.

161. Which of the following is most abundant in the earth crust

(a) Ga
(b) B
(c) Zn
(d) Al.

162. Carbon differs from other elements of the group. Which is the false statement?

(a) Due to d (-)orbitals in penultimate shell
(b) Due to it's marked tendency to from long chains catenation
(c) Due to it's limitation of coordination number 4
(d) Due to it's unique ablity to form long chains.

163. Chemically similar compounds differing by CH_2 group are called

(a) Allotropes
(b) Isomers
(c) Tautomers
(d) Homologouits.

164. Ammonical silver nitrate is used to distinguish between?
   (a) Acetylene and ethylene (b) Acetic acid and phenol
   (c) Acetone and ethnal (d) Ethylene and ethane.

165. which of the following is a true organometallie compound ?
   (a) Sodium methoxide (b) Sodium Amide
   (c) Sodium acetylide (d) Sodium carbonate.

166. Soft soaps are?
   (a) Sodium salts of fatty acids
   (b) Potassium salts of fatty acids containing excess of free alkali
   (c) Potassium salts of fatty acids containnig no free alkali
   (d) Calaium salts of fatty acids.

167. Ascorbic acids is?
   (a) Enzyme (b) Vitamin
   (c) Hormone (d) Proteins.

168. Who invented phonograph?
   (a) Edison (b) Bell
   (c) Grahm (d) Gouldestem.

169. Swedish chemist Karl Schelle obtained glycerine starting with an oil, Name the oil.
   (a) Olive oil (b) Middle oil
   (c) Light oil (d) fixed oil.

170. Name the largest silver producing country of the world.
   (a) India (b) Iran
   (c) Mexico (d) Canada.

171. Name an antibiotic drug that is prepared from mould
   (a) Heroine (b) Morphine
   (c) Penicillin (d) ISD.

172. Which compound's name has it's origin from a gum taken from plants growing near the temple of ammon?
   (a) Ammonium (b) Ammonia
   (c) Sodium (d) Coside.

173. Which type of charge is present on a proton?

(a) + tive (b) - tive
(c) No charge (d) Two positive.

174. By whom were X-rays discovered

(a) Curie (b) Rutherford
(c) Roentegen (d) Hann.

175. Choose the element that is used for preparing filaments of electric bulbs.

(a) Molybdenom (b) Nickel
(c) Tungsten (d) Chromium.

176. A nucleus of carbon is made up of x protons and y neutrons, give the numerical values of x and y.

(a) 6,6 (b) 6,7
(c) 7,6 (d) 6,8.

177. Name the metal that is used in storage of batteries

(a) Iron (b) Nickel
(c) Lead (d) Mercury.

178. What is the chemical formula of washing soda?

(a) NaCl (b) Na_2CO_3
(c) $NaHCo_3$ (d) Na_2SO_4.

179. What is the chemical formula of baking soda?

(a) KCl (b) Na_2CO_3
(c) $NaHCO_3$ (d) Na_2SO_4.

180. Baking power is a mixture of baking soda and another compound, name the other compound

(a) Citric acid (b) Tartaric acid
(c) Acetic acid (d) Oxalic acid.

181. Which compound is Known as caustic soda?

(a) NaOH (b) KOH
(c) $Na2CO_3$ (d) $NaHCO_3$.

182. Which compound is known as caustic potash?

(a) KOH (b) $K2CO_3$
(c) KNO_3 (d) K_2SO_4

183. Which compound is known as Lunar caustic?

(a) KOH (b) NaOH

(c) $AgNO_3$ (d) $PeSO_4$.

184. Which gas is generally filled up these days in electric bulbs to create an inert atmosphere?

(a) N_2 (b) He

(c) Ne (d) Ar.

185. Name the three kinds of rays given out by radioactive substances

(a) α-β-γ

(b) U.V.I.R X-rays

(c) Cathode rays, Anode rays α-rays

(d) Rontengen rays only.

186. Name the place where India's first indigeneous nuclear reactor is located

(a) Kota (b) Nangal

(c) Bombay (d) Trombay.

187. Name the sub atomic particle that is most likely to be displacted when we rub two objecs?

(a) Electron (b) Proton

(c) Meson (d) Neutron.

188. Name the gas that is common to all mineral acids

(a) Hydrogen (b) Oxygen

(c) Chlorine (d) Nitrogen.

189. Name the isotope of hydrogen present in heavy water.

(a) Deuterium (b) Tritium

(c) Protium (d) Proton.

190. Which year was radium discovered ?

(a) 1800 (b) 1850

(c) 1898 (d) 1793.

191. By whome was the existence of isotopes discovered

(a) Rutherford (b) Bhor

(c) Soddy (d) Smith.

192. In the treatment of Asthame, ... gases are used
   (a) Mixture of halium and oxygen
   (b) Mixture of neon and oxygen
   (c) Mixture of xenon and nitrogen
   (d) Mixture of argon and oxygen.

193. Moving neutrons by what name it is refered to?
   (a) Coolant (b) Moderator
   (c) Absorber (d) Accelerater.

194. By whom was the nuclear reactor invented?
   (a) Fergusen (b) Fermi
   (c) Frank (d) Bertholet.

195. Who was the incharge of USA atom bomb producing team?
   (a) Hann (b) Mattanhata
   (c) Openheimer (d) Anderson.

196. By whom was the periodic table of elements formulated ?
   (a) Newland (b) Dobernier
   (c) Mandeleev (d) Moseley.

197. What is the other name of Sulphuric acid ?
   (a) Oil of vitroil
   (b) Oil of Macrobine
   (c) Olive oil
   (d) Petroleum.

198. What is the name given to the device that is used for automatic control of temperature in furnaces?
   (a) Thermometer (b) Thermocol
   (c) Thermostat (d) Thermite.

199. Who discovered sodium (Na) ?
   (a) Davy (b) Dobernier
   (c) Stock holm (d) Schmidt.

200. Give the formula of Nitric acid
   (a) HCl (b) H_2SO_4
   (c) HNO_3 (d) NHO_2.

201. Who proved that air has weight ?

(a) Davy (b) Galileo
(c) Copernicus (d) Goldstein.

202. What is the atomic number of Oxygen?

(a) 8 (b) 16
(c) 32 (d) Zero.

203. Name the amino acid from which Urea is produced .

(a) Uric acid (b) Uranic acid
(c) Arginine (d) Digoxine.

204. Which is the most relative Halogen ?

(a) Fluorine (b) Iodine
(c) Chlorine (d) Bromine.

205. Which of the following elements has the smallest size?

(a) Magnesium (b) Phosphorus
(c) Chlorine (d) Aluminium.

206. One cc of any gas at N.T.P contains a definite number of molecules, give their number?

(a) 2.7×10^{12} (b) 2.7×10^{19}
(c) 6.02×10^{22} (d) $6.02 \times {}^{23}$.

207. If a gas obeys gas laws under all conditions then what do you call such a gas ?

(a) Ideal gas (b) Real gas
(c) Inert gas (d) Diatomic gas.

208. What is the source of sun's energy?

(a) Fusion
(b) Fission
(c) Partly
(d) Fusion and partly fission.

209. What is the name of the experiment by which the charge of an electron was determined for the first time ?

(a) Cathode (b) Ray experiment
(c) Decreases (d) Remains same.

210. Which of the following is the hardest metal ?

(a) Gold (b) Iron
(c) Platinum (d) Tungsten.

211. Which is the strongest acid ?

(a) F_2SO_4 (b) HCl
(c) HNO_3 (d) H_2CO_3.

212. Name the alkyne that can be obtained by Wulffs process.

(a) Methane (b) Ethane
(c) Ethylene (d) Acetylene.

213. Which of the following is a non-metal?

(a) Gallium (b) Indium
(c) Boron (d) Aluminium.

214. Which of the following is a polymer?

(a) Vinyl Chloride (b) Urea
(c) Starch (d) Styrene.

215. Name the alcoholic drink which is known as earn device

(a) Gin (b) Brandy
(c) Spirit (d) Beer.

216. A liquid called "coal oil" was discovered in 1856 name the liquid.

(a) Coal tar (b) Middle oil
(c) Petroleum (d) Kerosene oil.

217. Smog is

(a) Dew (b) polluted Air
(c) Condensed water (d) Fog.

218. CnH_2n+_2 is the general formula for

(a) Alkenes (b) Alkynes
(c) Alkanes (d) Cycloalkanes.

219. A saturated hydrocarbon is shown by CnH_{10}. The value of carbon atom 'n' in this compound is

(a) 2 (b) 5
(c) 6 (d) 4.

220. In alkanes the bond angle is

(a) 109.5° (b) 109°
(c) 120° (d) 180°.

221. The shape of methane molecule is

(a) linear (b) trigonal planer
(c) square planar (d) tetrahedral.

222. The approximate bond angle in ccly

(a) 190° (b) 109°
(c) 120° (d) 180°.

223. Which hydrocarbon is formed when aluminium carbide is reacted with aluminium carbide is reacted with water or dilute Hcl?

(a) Methane (b) Ethane
(c) Acetylene (d) Ethylene.

224. Which hydrocarbon is formed by the action of sodium on ethyl iodide ?

(a) Methane (b) Ethane
(c) Ethylene (d) Butane.

225. Both methane and ethane may be obtained in one step reaction from

(a) $C_2 H_4$ (b) CH_3l
(c) CH_3OH (d) C_2H_5OH.

226. When potassium acetate is electrolysed we get

(a) Methane (b) Ethane
(c) Acetylene (d) Ethylene.

227. CH_4 is formed when

(a) Sodium acetate is heated with soda lime
(b) Iodomethane is reduced
(c) Aluminium carbide reacts with water
(d) All.

228. Name the hydrocarbon that is a liquid at STP

(a) Ethane (b) Propone
(c) N-Butane (d) N-Pentane.

229. Which of the following liberates methane gas on treatment with water ?

(a) Silicon carbide (b) Calcium carbide
(c) Aluminium Carbide (d) Iron Carbide.

230. Methane can be prepared by

(a) Wurtz reaction
(b) Decarboxylation
(c) Hydrogenation reaction
(d) All.

231. Carbon black which is used in making printers ink is obtained by the decomposition of

(a) Acetylene (b) Benzene
(c) Carbon tetrachloride (d) Methane.

232. March gas mainly contains

(a) C_2H_2 (b) CH_4
(c) H_2S (d) CO.

233. Household gaseous fuel mainly contains

(a) CH_2 (b) C_2H_2
(c) C_2H_4 (d) C_4H_{10}.

234. General formula for alkenes is

(a) CnH2n (b) CnH2n+2
(c) C2nH2n (d) CnH2n-2.

235. Bond angle in alkenes is equal to

(a) 120° (b) 100° 28
(c) 180° (d) 60°.

236. Ethylene is a member of

(a) Alkanes (b) Olefins
(c) Alkynes (d) Paraffins.

237. General formula of alkynes is

(a) CnH2n (b) CnH2n+2
(c) C2nH2n (d) CnH2n-2.

238. % of hydrogen is maximum in
(a) CH_4 (b) C_2H_4
(c) C_6H_6 (d) C_2H_2.

239. General formula for benzene and its derivatives is
(a) CnH2n+6 (b) CnH2n+4
(c) CnH2n-6y (d) CnH2n+3.

240. Which of the following is a hydrocarbon
(a) Urea (b) Benzene
(c) Ammonium cyanate (d) Phenol.

241. Benezene was discovered by
(a) Ramsay (b) Dalton
(c) Faraday (d) Priestley.

242. The centric structure of benzene was proposed by
(a) Dewar (b) Ladenberg
(c) Kekule (d) Armstrong and Baeyer.

243. Carbon atoms in benzene molecule is inclined at an angle of
(a) 120° (b) 180°
(c) 109°28' (d) 60°.

244. Gammexene is
(a) D.D.T. (b) Benzene hexachloride
(c) Chloral (d) Hexachloroemethane.

245. Paraffin wax is
(a) Ester
(b) Alcohol
(c) Unsaturated hydrocarbon
(d) Saturated hydrocarbon.

246. Petroleum ether can be used as
(a) Solvent for fat, oil, varnish and rubber
(b) A fuel
(c) Both
(d) None.

247. Petrol is a mixture of
   (a) Alkenes (b) Alkanes
   (c) Alkynes (d) Aromatic hydrocarbons.

248. Kerosene oil is a mixture of
   (a) Alkanes (b) Aromatic compounds
   (c) Alkenes (d) Alkynes.

249. An octane number 100 is given to
   (a) N-hexane (b) Iso-octane
   (c) Neo-pentane (d) Neo-octane.

250. Kerosene oil is an
   (a) Aromatic compound (b) Gasoline
   (c) Hydrocarbon (d) Plant oil.

251. Structure of benzene is
   (a) Octahedral (b) Tetrahedral
   (c) Hexagonal (d) Triangular.

252. The six carbon-carbon bonds in benzene are
   (a) All equal to C=C bond length
   (b) All equal in bond length
   (c) Three of some and three of diefferent bond length
   (d) All different in bond length.

253. The main source of aromatic compounds is
   (a) Wood (b) Petroleum
   (c) Coal gas (d) Coal tar.

254. Liquid petroleum gas mostly contains
   (a) Methane (b) Ethane
   (c) Butane (d) Propane.

255. Hybridization of carbon in benzene is
   (a) Sp (b) Sp^2
   (c) Sp^3 (d) S^2p.

256. Which of the following is used as an anti-knocking material
   (a) TEL (b) C_2H_5OH
   (c) Glyoxal (d) Freon.

257. Which of the following bonds is strongest ?

(a) EC (b) C=C
(c) C ≡ C (d) None of these.

258. Compound formed when an alkylhalide reacts with magnesium is known as

(a) Alkane
(b) Benedict reagent
(c) Grignard reagent
(d) None.

259. A Grignard reagent may be made by reacting magnesium with

(a) Methyl amine (b) Diethylether
(c) Ethyl iodide (d) Ethyl alcohal.

260. Which is not present is Grignard reagent ?

(a) Methyl group (b) Magnesium
(c) Halogen (d) -COOH group.

261. Mark the wrong statement when chloroform is used as an anaesthetic the following precautions are observed

(a) It is kept in yellow bottles
(b) It is kept in open
(c) Bottles are filled upto the stopper
(d) One or two percent alcohal is added.

262. Chloroform can be used in medicine of

(a) Anaesthetic (b) Analgesic
(c) Antipyretic (d) Antibiotic.

263. Iodoform in medicine can be used as

(a) Anaesthetic (b) Antiseptic
(c) Analgesic (d) Antifebrin.

264. The refrigerant freon is

(a) Calcium flouride
(b) Hydroflu silicicacid
(c) Boron triflouride
(d) Dichlorodifluero-methane.

265. Which of the following compounds is used as a refrigerant ?
(a) Acetone (b) CCl_4
(c) CF_4 (d) CCl_2F_2.

266. Kerosene oil has compounds having
(a) C_5-C_8 (b) C_8-C_{12}
(c) C_{12}-C_{16} (d) C_{16}-C_{18}.

267. Gasoline is the name of
(a) Crude oil
(b) The gaseous constituents of petroleum
(c) The mixture of uncondensed gases produced in the distiliation of crude oil
(d) None of these.

268. Milled food products like atta are treated with which of the following ?
(a) BHC powder (b) Malathion
(c) DDT (d) Pyrethrum.

269. Match the following

I	DDT	(A) Fertilizer
II	Urea	(B) Pollutant
III	Sulfur dioxide	(C) Preservative
IV	Sodium benzoate	(D) Insecticide.

(a) I-A, II-B, III-C,IV-D
(b) I-D, II-A, III-B, IV-C
(c) I-D, II-A, III-C, IV-B
(d) I-B, II-C, III-A, I-C.

270. Dry ice is
(a) Carbon dioxide gas
(b) Solid carbon dioxide
(c) Mixture of air and water
(d) None of the above.

271. The sugar extracted from milk is known as
(a) lactose (b) Maltose
(c) Sucrose (d) Fructose.

272. The radiant energy of the sun is due to
   (a) Fusion (b) Fission
   (c) Convection (d) Conduction.

273. Which of the following is false?
   (a) Nickel tetra chloride is square planer in shape
   (b) Methane molecule is tetrahedral in shape
   (c) Acetylene is non-linear
   (d) P_2O_5 is likely two pyramider joined at their apices.

274. Maximum covalancy of an atom of an element is equal to ?
   (a) No of unpaired electrones in the p-orbitals
   (b) No of unpaired electrones in the s and p- orbitals
   (c) Total number of electrons in the s and p orbitals
   (d) Total number of electrons in the p orbitals.

275. Dalton's law of partial pressure is not applicable to
   (a) H_2 & Cl_{12} mixture (b) H_2 & CO_2 mixture
   (c) H_2 & N_2 mixture (d) All of these.

276. $[K_t+10 / K_3]$ is known as
   (a) Temeperature coefficient
   (b) Difference in temperature of reversible reactions
   (c) Ratio of equilibrium
   (d) None.

277. The passage of current throught a solution of certain electrolyte results in the liberation of H_2 at cathode & chlorine at anode. The electrolysis solutions ?
   (a) Water (b) NaCl in water
   (c) H_2SO_4 (d) $CuCl_2$ in water.

278. The conductance due to all the ions present in the solution of one gm equivalent of an electrolyte is ?
   (a) Specific conductance
   (b) Resistance
   (c) Molecular conductance
   (d) Equivalent conductance.

279. The structure of sodium chloride crystal is ?
   (a) Face centred cubic lattice
   (b) Square pianer
   (c) Octahedral
   (d) Body centred cubic lattice.

280. Tritium is
   (a) Active hydrogen
   (b) An isotope of hydrogen & deuterium
   (c) A mixture of hydrogen & deuterium
   (d) A compound of hydrogen.

281. German silver is made from
   (a) Cu, Cr, Sn (b) Ni, Zn, Cr
   (c) Cu, Al,Ni (d) Cu, Zn, Ni.

282. Portland cement is manufactured by using ?
   (a) Limestone, clay, sand
   (b) Limestone, clay, gypsum.
   (c) Limestone, gypsum, sand
   (d) Limestone, gypsum, alumina.

283. The valency of mercuric ion is ?
   (a) Two (b) Four
   (c) One (d) All.

284. Annydrous $AlCl_3$ is obtained from ?
   (a) HCl gas & Al metal (b) HCl & Al metal
   (c) Alumina & Cl_2 gas (d) None of these

285. Aluminium metal reduces many metalic oxides due to its great affinity for ?
   (a) Metals (b) Electrons
   (c) Hydrogen (d) Oxygen.

286. On strong heating lead nitrate gives?
   (a) PbO, NO, NO_2 (b) PbO_2, PbO, NO_2
   (c) PbO, NO_2, O_2 (d) PbO, NO,O_2.

287. Man dies when nitrotus oxide is inhaled in large quantity because ?
(a) It is poisounous
(b) It deactivates haemoglobin
(c) It causes laughing hysteria
(d) None of these.

288. Phosphorus is the name of ?
(a) Carbonyl chloride
(b) A phsphorus compound
(c) Phosphorus halide
(d) A phosphorium compound.

289. Dichlor aethylene shows ?
(a) Optical isomerism (b) Geometrical isomerism
(c) Both a and b (d) None of the above.

290. Halogenation of propane is ?
(a) Electrophilic reaction (b) Free radical reaction
(c) Ionic reaction (d) Nucleophilic reaction.

291. Methylated spirit is ?
(a) Methanal and traces of water
(b) Ethanel with traces of water
(c) Ethanal containing methanal
(d) Only methanal.

292. Reaction of alkyle halide with sodium alkoxide gives ?
(a) Ketone (b) Ether
(c) Carbooxylic acid (d) Alcohol.

293. Which of the following compound is oxidised to propane methyle ethyl ketone?
(a) 1-Butanal (b) 2- Propanal
(c) 2- Butanal (d) 1- Butyl alcohol.

294. Amorphoes substances possess an atomic structure in which geometric arrangement?
(a) Very regular (b) Well defined
(c) Varies (d) Remains same.

295. An atom containing an old number of electrons is
   (a) Paramagnetic (b) Diamagnetic
   (c) Ferromagnetic (d) None of these.

296. An element is best characterised by its
   (a) Mass number (b) Atomic number
   (c) Atomic volume (d) Oiling & melting points.

297. To which famiiy does 'As' belong ?
   (a) Oxygen (b) Hlogen
   (c) Boron (d) Nitrogen.

298. Transition metals
   (a) Exhibit diamagnetism
   (b) Under to inert pair affect
   (c) Do not from alloy
   (d) Show variable oxidation states.

299. The total number of electrons in energy level 'n' is
   (a) N (b) $2n^2$
   (c) $N^2/_2$ (d) $(2n)^2$.

300. The element with highest electron affinity in
   (a) O (b) Cl
   (c) N (d) F.

301. An electron emmited from the nucleus of a radioactive element is called ?
   (a) A positron (b) Anti-proton
   (c) Beta-particle (d) None of these.

302. Correct gas equation is
   (a) $\frac{P_1V_1}{P_2V_2} = \frac{T_1}{T_2}$ (b) $\frac{V_1T_2}{P_1} = \frac{P_2T_1}{T_2}$
   (c) $\frac{P_1V_1}{V_1} = \frac{P_2T_2}{V_2}$ (d) $\frac{V_1V_2}{T_1T_2} = P_1V_2$.

303. The size of particles in suspension solution and Colloide varies the order

(a) Suspension > solution > colloid
(b) Solution > colloid > solution
(c) Solution > colloid > suspension
(d) Colloid > solution > suspension.

304. Random motion of a particle is known as

(a) Tyndall effect
(b) Dialysis
(c) Brownion movement
(d) Electrosmosis.

305. The sky looked blue due to ---------------- of light

(a) Dispersion (b) Reflection
(c) Transmission (d) Scattering.

306. Which of the following is a lyophobic sol?

(a) Gelatin (b) Sulphur
(c) Gum arabic (d) Starch.

307. Least chemical activity is shown by

(a) Ammonia (b) Methane
(c) Argon (d) Sulphuric acid.

308. In all their compounds alkali metals are

(a) Monovalent (b) Polyvalent
(c) Divalent (d) Zero valent.

309. Gold is soluble in

(a) HCl
(b) HNO_3
(c) H_2SO_4 saturated in SO_3
(d) 3 parts of HCl + 1 part of HNO_3.

310. Parkes process is used in the extraction of

(a) Ag (b) Cu
(c) Zn (d) Na.

311. Essential constituent of amalgam is

(a) Iron (b) An alkali metal
(c) Silver (d) Mercury.

312. The element which forms ions in dimeric state is

(a) Mercury (b) Cadmium
(c) Calcium (d) Zinc.

313. Meatal present in chalophyll is

(a) Ca (b) Mg
(c) Zn (d) Cd.

314. Plaster of Paris

(a) Expands while setting
(b) Contracts while setting
(c) Undergoes no charge while setting
(d) Reduce to half of the initial volume.

315. Milk of lime is

(a) MgO (b) CaO
(c) $Ca(OH)_2$ (d) $CaCO_3$.

316. Of the Compound the most acidic is

(a) AS_2O_3 (b) P_2O_5
(c) Sb_2O_3 (d) Bi_2O_2.

317. Of the different isotopes of phosphorus the most reactive is F?

(a) Violet (b) White
(c) Black (d) Scarletred.

318. A lighting flash in air may result in the formation of

(a) Nitrogen pentoxide (b) Nitrous oxide
(c) Nitric oxide (d) Active nitrogen.

Which of the following show highest tendency of calcination

(a) Oxygen (b) Sulphur
(c) Selenium (d) Tellurium.

320. Mark the smallest atom

(a) Fluorine (b) Chlorine
(c) Bromine (d) lodine.

321. The main source of most of organic compound is

(a) Coaltar (b) Petroleum
(c) A and b (d) None of the above.

322. Petrol is a mixture of

(a) Alkenes (b) Alkanes
(c) Alkynes (d) Aromatic hydrocarbons.

323. Radioactivity was discovered by

(a) Henry Becquerel (b) Rutherford
(c) J.J. Thomson (d) Madame Curie.

324. Radioactive decay is

(a) A physical change (b) A chemical change
(c) A nuclear change (d) None of these.

325. In terms of energy I.a.m.u. is equal to

(a) 100 J (b) 931.1 mev
(c) 931.1k cal (d) 10^7 erg.

326. Choose house the element which is not radioactive ?

(a) Cm (b) No
(c) Mo (d) Md.

327. Which of the following is radioactive element?

(a) Sulpher (b) Polonium
(c) Tellurium (d) Selenium.

328. Radioactive is due to

(a) Stable electronic configuration
(b) Unstable electronic configuration
(c) Stable nucleus
(d) Unstable nucleus.

329. The electron is

(a) An α particle (b) A β-ray particle
(c) A hydrogen ion (d) A position.

330. Gamma rays are
   (a) High energy electron
   (b) Low energy electron
   (c) High energy electromagnetic waves
   (d) High energy positrons.

331. Heaviest particle is
   (a) Meson (b) Neutron
   (c) Protone (d) Electron.

332. Positron is
   (a) Electron positive charge
   (b) Helium nucleus
   (c) A nucleus with two protons
   (d) A nucleus with one neutron and proton.

333. In radioactive decay which one of the following moves the faster ?
   (a) A- particles (b) B- particles
   (c) Γ-rays (d) Positron.

334. Penetrating power of α-particles is
   (a) More than γ-rays (b) More than β-rays
   (c) Less than B-rays (d) Nil.

335. The rays which are deflected in a magnetic field
   (a) Only α-rays (b) Only γ-rays
   (c) A-and β-rays (d) B-rays and γ-rays.

336. Which is correct statement ?
   (a) Isotopes are always radioactive
   (b) B rays are always negatively charged particles
   (c) A rays are always negatively charged particles
   (d) Γ rays can be deflected in a magnetic field.

337. $_{92}U^{238}$ emits an particle the product has mass number and atomic number
   (a) 236, 92 (b) 234, 90
   (c) 238, 90 (d) 236, 90.

338. Loss of an article is equivalent to
(a) Loss of two neutrons
(b) Loss of two protons
(c) Loss of two neutrons and loss of two protons
(d) None of the above.

339. The group displacement law was given by
(a) Bacquerell (b) Rutherford
(c) Soddy and Fajan (d) Mendeleef.

340. Complete the reaction

$_7N^{14} + {}_0n^1 \text{-----}> {}_{+1}H^1$

(a) $_6C^{14}$ (b) $_6C^{13}$
(c) $_6C^{12}$ (d) $_7N^{15}$.

341. Who gave the theory of radioactive disintegration ?
(a) Rutherford and Soddy (b) Soddy and Thomson
(c) Rutherford (d) Soddy.

342. Which of the following does not have neutron in its nucleus ?
(a) Hydrogen (b) Deuerium
(c) Tritium (d) None.

343. $^{40}_{20}$Ca and $^{40}_{19}$K are
(a) Isomers (b) Isotopes
(c) Isobars (d) Isoelectronic.

344. Difference between N^{20} and N^{22} is
(a) 12 Protons and 2 neutrons
(b) 10 neutrons and 12 proton
(c) 2 neutron and 2 electron
(d) 2 neutrons only.

345. The fuel used in nuclear reactor is
(a) Heavy water (b) Graphite
(c) Lead (d) Uranium.

346. Which of the following nuclei is unstable?

(a) ${}_5B^{10}$ (b) ${}_4B^{10}$

(c) ${}_7N^{14}$ (d) ${}_8O^{16}$.

347. The end product formed by the disntegration of $^{222}_{86}$Ra is

(a) $^{204}_{81}$Ti (b) $^{206}_{82}$Pb

(c) $^{222}_{86}$Ra (d) $^{207}_{83}$Bi.

348. Fission reaction takes place in

(a) A nucleur reactor only

(b) An atomic bomb only

(c) A reactor as well as an atomic bomb

(d) The sun.

349. Heavenly matter landing on the earth surface is known as

(a) Meteor (b) Shooting star

(c) Meteorite (d) UEO.

350. A saturated solution at a particular temperature

(a) Is colourless

(b) Is highly coloured

(c) Does not dissolve any more of the solid solute.

(d) Contains suspended solid particles.

351. The weight of a rusted iron object is that of an unrusted one.

(a) Appreciately more than (b) The same as

(c) Less than (d) More or less the some as.

352. A gas which is not present in normal samples of air is

(a) Neon

(b) Chlorine

(c) Carbondioxide

(d) Helium.

353. The substance that is added to made natural ruther strong and more bouncy is

(a) Sulphuric acid (b) Sponge
(c) Sulphur (d) Chlorine.

354. The substance made of three carbon is

(a) Sugar (b) Acetic acid
(c) Graphite (d) Methane.

355. Fermentation of glucose finally gives

(a) CO_2 and CH_3 OH
(b) CO and alcohol
(c) CO_2 and H_2O
(d) CO_2 and C_2H_5 OH.

356. The element which forms the largest number of compounds with hydrogen is

(a) Oxygen (b) Silicon
(c) Carbon (d) Boron.

357. Chloromyce tin is

(a) An anticeptic (b) Antidepressed
(c) Analgesic (d) Antibacterial.

358. The acid that can be used as a hypnotic is

(a) Tartaric acid (b) Benzoic acid
(c) Barbitaric acid (d) Butanic acid.

359. Which of the following substances, the one that can be used as an insecticide is-

(a) DDT (b) Gunpowder
(c) TNT (d) Paracetamol.

360. Acetyl salicyclic acid is commonly used as-

(a) Tear gas (b) A fertiliser
(c) A pain killer (d) A sedotine.

361. The gas in atmospheric air that helps in burning is

(a) Nitrogen (b) Helium
(c) Oxygen (d) Carbon dioxcide.

362. Use of gypsym is recommended mainly in soils which on-

(a) Alkaline (b) Saline

(c) Water-logged (d) Acidic.

363. Anhydrouas calcium chloride acts as a/an

(a) Dehydrating agent (b) Drug

(c) Oxidant (d) Mordant.

364. Cheomotherapy deals with

(a) Industrial engineering.

(b) The use of chemicals in warfare.

(c) The study and uses of chemicals In the cure of diseases.

(d) The use of chemicals in the food industry.

365. Inert gases are-

(a) Miscible with water (b) Not stable

(c) Chemically unreactive (d) Chemically very active.

366. The branch of chemistry dealing with the accurate determination of the amounts of various substances is called

(a) Biochemistry

(b) Inorganic chemistry

(c) Organo metallic chemistry

(d) Analytical chemistry.

367. Alloy steel cotaining chormium to resist rusting is known as

(a) Wrought iron (b) Cast iron

(c) Hard iron (d) Stainless steel.

368. Petroleum is found in

(a) Igneous rocks (b) Sedimentary rocks

(c) Metamorphic rocks (d) Marshy land.

369. Allugel tablets used to reduce acidity contain

(a) Sodium carbonate

(b) Sodium hydroxide

(c) Magnesium hydroxide

(d) Aluminium hydroxide.

370. The drug against infectious diseases is
(a) Suphathiazale (b) Insulin
(c) Aspirin (d) Resenpine.

371. Indigo is
(a) Vate dye (b) Basic dye
(c) Acidic dye (d) Ingrain dye.

372. Amount of N_2 in air by volume is about
(a) 50% (b) 80%
(c) 20% (d) 25%.

373. Element that fails to form compound is
(a) P (b) N
(c) Sb (d) Ba.

374. The percentage of O_2 by Volume in the air is
(a) 20.99 (b) 23.00
(c) 29.50 (d) 81.82.

375. The most abundant of the rare gases in air is
(a) Neon (b) Argon
(c) Krypton (d) Helium.

376. Which of the following is not an inert gas ?
(a) Argon (b) Xenon
(c) Radon (d) Fluorine.

377. Hard water
(a) Gives good lather with soap
(b) Does not give good lather with soap
(c) Gives good lather with soap and oil
(d) Does not give good lather with cream.

378. Temporary hardness of water is due to presence of
(a) Chloride of Ca and Mg
(b) Nitrates of Ca and Mg
(c) Bicarbonates of Ca and Mg
(d) Sulphates of Ca and Mg.

379. Water has high boiling point because
   (a) Its molecules are associated by hydrogen bonds
   (b) It is mixture
   (c) It is liquid
   (d) It is mixture of O_2 and H_2.

380. Water is good solvent because
   (a) It is liquid form
   (b) Its boling point is very high
   (c) Its molecules are highly polar
   (d) It is in dilute form.

381. Pure water is a bad conductor of electricity because it is
   (a) Only feebly ionised
   (b) A very good solvent
   (c) Highly associated by hydrogen bonds
   (d) Appereciable volatile.

382. Pure water does not react with
   (a) Calcium carbonate
   (b) Calcium nitride
   (c) Calcium hydride
   (d) Calcium carbide.

383. The first element of the periodic table is
   (a) Sodium (b) Neon
   (c) Hydrogen (d) Helium.

384. Number of isotopes of hydrogen is
   (a) 1 (b) 2
   (c) 3 (d) 4.

385. During day time plants absorb
   (a) Oxygen
   (b) Nitrogen
   (c) Combined mixture of O_2 and CO_2 gases
   (d) Carbon dioxide.

386. Hydrogen peroxide is
   (a) An oxidising agent
   (b) Both oxidising and reducing agent
   (c) A reducing agent
   (d) None of the above.

387. Nature of sodium carbonate solution is
   (a) Basic (b) Acide
   (c) Neutral (d) None of these.

388. Carbon is a
   (a) Metal (b) Non-metal
   (c) Metalloid (d) None of these.

389. The hardest variety of carbon is
   (a) Coke (b) Gas carbon
   (c) Diamond (d) Graphite.

390. Which of the followng is not an allotrophic form of carbon?
   (a) Diamond (b) Graphite
   (c) Amorphous carbon (d) Carborandum.

391. The form of carbon known as charcoal
   (a) Is amorphous
   (b) Consists of micro-crystals of graphite
   (c) Consists of micro-crystals of diamond
   (d) Is lighter than water.

392. Which of the following is not an amorphous form of carbon
   (a) Coke (b) Charcoal
   (c) Lamp black (d) Graphite.

393. Air becomes incombustible if its carbon dioxide content is
   (a) 15% (b) 2.5%
   (c) 7.5% (d) 5%.

394. Which of the following is not a natural fertilizer
   (a) Bone meal (b) Chile saltpetre
   (c) Urea (d) Dried blood.

395. The number of electrons in the outermost shell of halogen is
(a) 7 (b) 5
(c) 3 (d) None of these.

396. The compound that shows electrovalent linkage is
(a) $ChCl_3$ (b) CCl_4
(c) NaCl (d) Ch_3Cl.

397. Nitrogen is
(a) Lighter than air (b) Lighter than NH_3
(c) Heavier than air (d) Heavier than nitric oxide.

398. Dense white fumes result when ammonia in contact with
(a) HNO_3 (b) HCl
(c) H_2SO_4 (d) NaCl.

399. Calgon is used as a
(a) Fertilizer (b) Water softener
(c) Insecticide (d) Pesticide.

400. The halogen that is liquid at room temperature is
(a) Fluorine (b) Chlorine
(c) Bromine (d) Pesticide.

ANSWER SHEET

1.	(a)	2.	(c)	3.	(b)	4.	(d)
5.	(a)	6.	(b)	7.	(b)	8.	(b)
9.	(a)	10.	(b)	11.	(d)	12.	(b)
13.	(a)	14.	(a)	15.	(a)	16.	(a)
17.	(b)	18.	(c)	19.	(c)	20.	(c)
21.	(c)	22.	(a)	23.	(a)	24.	(a)
25.	(c)	26.	(a)	27.	(a)	28.	(b)
29.	(a)	30.	(a)	31.	(b)	32.	(c)
33.	(c)	34.	(b)	35.	(d)	36.	(b)
37.	(a)	38.	(b)	39.	(b)	40.	(c)
41.	(b)	42.	(d)	43.	(a)	44.	(a)
45.	(a)	46.	(a)	47.	(b)	48.	(b)

49.	(a)	50.	(d)	51.	(d)	52.	(b)
53.	(c)	54.	(a)	55.	(b)	56.	(c)
57.	(a)	58.	(c)	59.	(c)	60.	(d)
61.	(b)	62.	(a)	63.	(b)	64.	(c)
65.	(d)	66.	(b)	67.	(d)	68.	(a)
69.	(b)	70.	(a)	71.	(b)	72.	(d)
73.	(a)	74.	(d)	75.	(a)	76.	(a)
77.	(d)	78.	(a)	79.	(a)	80.	(d)
81.	(a)	82.	(a)	83.	(d)	84.	(c)
85.	(b)	86.	(c)	87.	(c)	88.	(b)
89.	(a)	90.	(b)	91.	(b)	92.	(c)
93.	(a)	94.	(a)	95.	(a)	96.	(c)
97.	(d)	98.	(b)	99.	(c)	100.	(b)
101.	(a)	102.	(b)	103.	(a)	104.	(c)
105.	(c)	106.	(b)	107.	(c)	108.	(b)
109.	(b)	110.	(a)	111.	(b)	112.	(a)
113.	(b)	114.	(d)	115.	(c)	116.	(a)
117.	(d)	118.	(d)	119.	(c)	120.	(b)
121.	(c)	122.	(b)	123.	(b)	124.	(b)
125.	(c)	126.	(a)	127.	(b)	128.	(c)
129.	(a)	130.	(a)	131.	(d)	132.	(d)
133.	(c)	134.	(c)	135.	(b)	136.	(c)
137.	(c)	138.	(b)	139.	(d)	140.	(d)
141.	(b)	142.	(a)	143.	(c)	144.	(c)
145.	(b)	146.	(a)	147.	(c)	148.	(c)
149.	(d)	150.	(c)	151.	(a)	152.	(d)
153.	(d)	154.	(c)	155.	(c)	156.	(b)
157.	(a)	158.	(b)	159.	(c)	160.	(d)
161.	(d)	162.	(a)	163.	(d)	164.	(a)
165.	(c)	166.	(c)	167.	(b)	168.	(a)
169.	(a)	170.	(c)	171.	(c)	172.	(b)
173.	(a)	174.	(c)	175.	(c)	176.	(a)
177.	(c)	178.	(d)	179.	(c)	180.	(d)
181.	(a)	182.	(a)	183.	(c)	184.	(d)
185.	(a)	186.	(b)	187.	(a)	188.	(a)
189.	(a)	190.	(c)	191.	(c)	192.	(a)
193.	(b)	194.	(c)	195.	(c)	196.	(c)
197.	(a)	198.	(c)	199.	(a)	200.	(c)
201.	(b)	202.	(a)	203.	(c)	204.	(a)
205.	(a)	206.	(a)	207.	(a)	208.	(a)

209. (b)	210. (d)	211. (a)	212. (d)
213. (c)	214. (c)	215. (b)	216. (c)
217. (c)	218. (c)	219. (d)	220. (a)
221. (d)	222. (b)	223. (a)	224. (d)
225. (b)	226. (d)	227. (d)	228. (d)
229. (c)	230. (b)	231. (d)	232. (b)
233. (d)	234. (a)	235. (a)	236. (b)
237. (d)	238. (a)	239. (c)	240. (b)
241. (c)	242. (d)	243. (a)	244. (b)
245. (d)	246. (a)	247. (b)	248. (a)
249. (b)	250. (c)	251. (c)	252. (b)
253. (d)	254. (c)	255. (b)	256. (a)
257. (d)	258. (c)	259. (c)	260. (d)
261. (b)	262. (a)	263. (b)	264. (d)
265. (d)	266. (c)	267. (b)	268. (a)
269. (b)	270. (b)	271. (a)	272. (a)
273. (c)	274. (a)	275. (a)	276. (a)
277. (b)	278. (d)	279. (a)	280. (b)
281. (d)	282. (b)	283. (a)	284. (d)
285. (c)	286. (c)	287. (a)	288. (b)
289. (b)	290. (b)	291. (b)	292. (b)
293. (c)	294. (c)	295. (a)	296. (b)
297. (a)	298. (d)	299. (b)	300. (b)
301. (c)	302. (a)	303. (b)	304. (c)
305. (d)	306. (b)	307. (c)	308. (a)
309. (d)	310. (a)	311. (c)	312. (a)
313. (b)	314. (a)	315. (b)	316. (b)
317. (b)	318. (c)	319. (b)	320. (a)
321. (c)	322. (c)	323. (a)	324. (c)
325. (b)	326. (c)	327. (b)	328. (d)
329. (b)	330. (c)	331. (c)	332. (a)
333. (c)	334. (c)	335. (c)	336. (b)
337. (b)	338. (c)	339. (c)	340. (a)
341. (a)	342. (a)	343. (c)	344. (d)
345. (d)	346. (b)	347. (b)	348. (c)
349. (c)	350. (c)	351. (a)	352. (c)
353. (c)	354. (c)	355. (c)	356. (c)
357. (d)	358. (c)	359. (a)	360. (c)
361. (c)	362. (a)	363. (a)	364. (c)
365. (c)	366. (d)	367. (d)	368. (a)
369. (d)	370. (a)	371. (d)	372. (b)

373.	(c)	374.	(a)	375.	(b)	376.	(d)
377.	(b)	378.	(c)	379.	(a)	380.	(c)
381.	(a)	382.	(b)	383.	(c)	384.	(c)
385.	(d)	386.	(b)	387.	(a)	388.	(d)
389.	(c)	390.	(d)	391.	(a)	392.	(d)
393.	(b)	394.	(c)	395.	(a)	396.	(c)
397.	(a)	398.	(b)	399.	(b)	400.	(a)

5
SCIENTIFIC ACHIEVEMENTS DURING LAST FIVE DECADES

1. Science Day is celebrated on February 28 every year because:

   (a) Raman effect came to light this day Feb 28.

   (b) Raman got Nobel Prize

   (c) Raman's date of birty

   (d) Date of birth of J.C.Basu.

2. Match the names of Indian Scientist in table II and Strike down the following correct Sequence:Table I

   (a) Dr. Raja Ramanna

   (b) Dr. M.S. Swaminathan

   (c) Prof. U.R. Rao

   (d) Prof. Meghnad Saha.

   1. Botanical chemistry
   2. Nucleary Physics
   3. Heat, dynamics & Astrophysics
   4. Space Research
   5. Agricultural Scince.

	A	B	C	D.
(a)	3	5	2	1
(b)	2	1	4	3
(c)	2	5	4	3
(d)	3	1	4	2

3. Which of the followings is not properly matched
   (a) Sushrut - Surgery
   (b) Charak - Medicinal treatment & Physician
   (c) Aryabhatta - Metallurgy
   (d) Bhaskar - Principal of Algebra.

4. Asiatic Society was established by
   (a) Dr. Mahendra Lal Sarkar
   (b) Dr. V. R. Ambedkar
   (c) Dr. William Jones
   (d) Dr. William Jones.

5. The Indian Association for the cultivation of Science was formed by
   (a) Dr. Mahendra Lal Sarkar
   (b) Dr. V.R. Ambedkar
   (c) Dr. William Jones.

6. 'Raman Research Institute' is a part of-
   (a) Asian Society
   (b) Indian Science Congress Association
   (c) Bose Institute
   (d) Indian Science Academy.

7. Name who propounded theory of numbers:
   (a) S.A. Ramanujam
   (b) Satyendra Nath Basu
   (c) Jagdish Chandra Basu
   (d) Meghnad Saha.

8. Who was a famous Chemist?
   (a) S.A.Ramannujan (b) Satyendra Nath Bose
   (c) Jagdish Chandra Basu (d) Meghnad Saha.

9. Who is credited for forming Indian Science Congress Association
   (a) Sir C.V. Raman (b) Satyendra Nath Basu
   (c) Birbal Sahni (d) Prafulla Chandra Roy.

10. Put Indian Science Congress Association, Indian Congress Academy, Science and Technological Research Council, Indian Science Institution all in chronological order by coding them serially as A, B, C, & D.

(a) A, B, C, D (b) D, A,B, C,
(c) C, D, B, A (d) B, A, D, C.

11. Indian Science Institute and Bose Institute is situated respectively at :

(a) Bombay & Calcutta (b) Bangalore & Calcutta
(c) Bangalore & Bombay (d) Bombay & Bangalore.

12. The main purpose behind establishing Asiatic Society was:

(a) To accelerate the pace of industrial development in India
(b) To arrange an enriched laboratory for research work
(c) To print the thesis papers of Physics, Chemistry, Geography etc.
(d) All of these.

13. Who is named as the 'Father of Modern Indian Science?

(a) Pt. Jawaharlal Nehru
(b) Smt. Indira Gandhi
(c) Dr. Rajendra Prasad
(d) Late. Mahatma Gandhi.

14. Which one among the following is mis-matched?

(a) National Chemical Laboratory, Pune
(b) National Physical Laboratory, Pune
(c) National Botanical Garden, Lucknow
(d) Central Medicine Research Institute, Delhi.

15. National Chemical Laboratory is situated at:

(a) Bombay (b) Pune
(c) Mysore (d) Delhi.

16 Central Fuel Research Institute is situated at:

(a) Delhi (b) Jadavpur
(c) Dhanbad (d) Pune.

17. Which one of the following is not matching?
    (a) Central Building Research Institute, Roorkee
    (b) Central Road Research Institute, Delhi
    (c) National Metalleogical Laboratory, Pilani
    (d) Central Leather Research Institute, Chennai.

18. The first Chairman of qutanomous CSIR
    (a) Dr. Homi Jahangir Bhabha
    (b) Dr. Vikram Sarabhai
    (c) Dr. Shanti Swarup Bhatnagar
    (d) None of these.

19. The total no of laborataries & research institutions under CSIR are:
    (a) 14 (b) 24
    (c) 30 (d) 40.

20. The Scientific Research and Natural Resources Ministry was established in India in the year:
    (a) 1931 (b) 1941
    (c) 1951 (d) 1961.

21. The resolution for National Science policies was passed on:
    (a) 30 January, 1950 (b) 4 March, 1957
    (c) 4 March, 1958 (d) 30 January, 1958.

22. According to the National Science policies, the factors needed for the prosperity of the Nation are:
    (a) Raw material , Energy , Technology
    (b) Raw material , Capital , Technology
    (c) Energy , Capital , Raw, material
    (d) Energy , Technology & Raw material.

23. Which one of the following is not included in the objectives on the National Science Policy:
    (a) To encourage creative talents
    (b) To ensure availability Suprior Research Scientist
    (c) Scientists be trained to the highest level.
    (d) To encourage the development of non-conventional source of energy.

24. Who announced Technological policies?
    (a) Pt. Jawahar Lal Nehru
    (b) Smt. Indira Gandhi
    (c) Shri. Rajiv Gandhi
    (d) Shri. P. V. Narasimha Rao.

25. Which one the following was the part of National Technological policies:
    (a) Recycling of waste-materials
    (b) Encouraging the non-conventional source of energy.
    (c) Developing benefitial technologies of international standard for the export.
    (d) All of these.

26. National Technological policy was announced on:
    (a) 30 January, 1983 (b) 3 January, 1983
    (c) 23 January, 1983 (d) 30 January, 1950.

27. Which institution is related together the informations on technological research being held inside and outside of the country and to make it available to related areas:
    (a) CSIR (b) DRDO
    (c) TFRI (d) TIFAC.

28. New Technological policy was announced on:
    (a) 1953 (b) 1983
    (c) 1988 (d) 1993.

29. The Stream was put in the New Technological Policy on:
    (a) Quality of international Standard
    (b) Use of traditional handicraft
    (c) development of technology for expert
    (d) All of these.

30. How many departments are functioning under Central Government in India. Which are related to the activities of Science and Technology:
    (a) Four (b) Five
    (c) Six (d) Seven.

31. 'Operation Shakti' is the code word for:
   (a) Pokhran atomic explosion expedition
   (b) War- exercise by S.U. 30 aeroplane of Indian Air force.
   (c) Preventive programme to stop terrorism by the Home Minister of India in J&K
   (d) Military action against terrorist of Tripura by Indian Armed Force.

32. The difference between a nuclear reactor and an atom bomb is that:
   (a) There is no serise Processing in a nuclear reactor, whereas an atom bomb is having it:
   (b) In nuclear reactor processing of series is kept under control.
   (c) In nuclear reactor processing of series is not under control.
   (d) An atom-bomb is based on nuclear fusion, whereas nuclear-reactor is based on nuclear- fission.

33. Which one of the following is used as a cooling agent in the 'Fast breeder Test Reactor of Kalapakkam?
   (a) Carbon-di- oxide (b) Heavy water
   (c) Oceanic water (d) Melted Sodium.

34. Where in Uttar Pradesh, two Atomic Energy Units' are being set up?
   (a) Haridwar (b) Mussorie
   (c) Naraura (d) Narendranagar.

35. With which field CTBT is related to?
   (a) Pact on preventing tests of atomic armaments
   (b) International Sports
   (c) Animals trade
   (d) Prevention of child-labours.

36. In May 1998, India carried tests of five atomic explosives in Pokhran? Can you tell in which State, Pokhran is located?
   (a) Gujarat (b) Rajasthan
   (c) Punjab (d) Himachal Pradesh.

37. Who led the team of scientists, that made atomic tests in may 1998 in India?
    (a) Abdul Hashim (b) Md. Usman
    (c) A.P. J. Abdul Kalam (d) Abdul Kasim.

38. India Carried second atomic tests on
    (a) May 10,1998 (b) May 11,1998
    (c) May 12, 1998 (d) May 16, 1998.

39. Tata Fundamental Research Institute was set up in the year.
    (a) 1944 (b) 1945
    (c) 1947 (d) 1948

40. Name by whom investment was made for Tata Fundamental Research Institute?
    (a) British Government (b) Dadabhai Nawroji Trust
    (c) America (d) Darabji Trust.

41. Atomic Energy Comission was set up in the year:
    (a) 1944 (b) 1945
    (c) 1947 (d) 1947.

42. He was chairman at the time of organisation of Atomic Energy Commission:
    (a) Dr. Shanti Swarwp Bhatnagar
    (b) Dr. Rajendra Prasad
    (c) Dr. H. J. Bhabha
    (d) Pt. Jawhar Lal Nehru.

43. Atomic Energy law was formulated in the year:
    (a) 1944 (b) 1947
    (c) 1948 (d) 1949.

44. 'Scarce Minerals Survey Unit' was set up in the year place:
    (a) Delhi, 1948 (b) Bombay, 1948
    (c) Delhi, 1949 (d) Bombay, 1949.

45. Departmant of Atomic Energy was set up in the year
    (a) 1948 (b) 1950
    (c) 1956 (d) 1962.

46. Who of the following handles the Department of Atomic Energy
    (a) Defence minister
    (b) Energy minister
    (c) Prime Minister
    (d) President.

47. Which one of the following is not the aim of the Department of Atomic Energy?
    (a) Safe production of Electricity
    (b) Setting up of nuclear reactor from nuclear energy
    (c) Encourage Technology exchange
    (d) None of these.

48. Which one of the following institutions is under the Department of Atomic Energy?
    (a) Advance Industrial Centre, Indore
    (b) Heavy water Board, Bombay
    (c) Nuclear Fuel Composition, Hyderabad
    (d) Indian Rare Earths Ltd, Bombay.

49. Bhabha Atomic Research Centre is situated at:
    (a) Delhi
    (b) Hyderabad
    (c) Trombay
    (d) Calcutta.

50. Which one of the following institutions is situated at Kalpakkam?
    (a) Advance Technology Centre
    (b) Department of Atomic Energy
    (c) Indira Gandhi Atomic Research Centre
    (d) Atomic Energy Educational Centre.

51. Find out the factual one of the following:
    (a) India need not require 'heavy water'
    (b) India imports 'Heavy water'
    (c) India is self- department for 'Heavy water'
    (d) India is an exporting country of 'Heavy water'.

52. The weight of the atom-bomb exploded in the year 1947 was
    (a) 1.2 tonnes
    (b) 12 tonnes
    (c) 1.2 kilotonnes
    (d) 12 Kilotonnes.

53. The five tests commenced in Pokhran in the year 1998 were of the device

(a) One thermonuclear, one fission, three nuclear
(b) One thermonuclear, two fission, two nuclear
(c) One thermonuclear, three fission, one nuclear
(d) One thermonuclear, four fission.

54. The 'Hydrogen bomb' tested in the year 1998 by India was of:

(a) 43 Kilotonnes (b) 15 Kilotonnes
(c) 0.2 Kilotonnes (d) 1.2 Kilotonnes

55. The test explosion on May 13, 1998 was of

(a) 0.2 kilotonnes (b) 0.3 kilotonnes
(c) 0.5 kilotonnes (d) 0.3 & 0.5 kilotonnes

56. India has occupied which one of the following place and Pokharan test:

(a) Fifth (b) Sixth
(c) Seventh (d) Eight.

57. Who is credited with laying foundation stone of Nuclear Energy Programme in India?

(a) Dr. Abdul Kalam (b) Dr. Vikaram Sarabhai
(c) Dr. Bhabha (d) Dr. S.S Bhatnagar.

58. How much quantity of Uranium can produce the same amount of energy, which we could get by burning 20 tonnes coal?

(a) 0.1 tonnes (b) 0.2 tonnes
(c) 0.5 tonnes (d) 1 tonn.

59. Which of the following minral is found in enough quantity in India?

(a) Uranium 235 (b) Uranium 233
(c) Plutonium (d) Thorium.

60. 'Atomic Energy Programme' of India was divided into how many stages by Dr. Bhabha?

(a) Two (b) Three
(c) Four (d) Five.

61. In the first stage of 'Atomic Energy programme', the type of reactor, which was made was:
    (a) To use Uranium as fuel and produce Plutonium
    (b) To use Uranium as fuel and produce Thorium
    (c) To use plutonium as fuel and produce Thorium
    (d) To produce thorium.

62. In the second stage of 'Atomic Energy Programme' the fuel used by the Fast Breeder Reactar is.
    (a) Uranium
    (b) Plutonium
    (c) Thorium
    (d) Molibdenum.

63. In the first stage of 'Atomic Energy Programme' the Pressurised Hevy water Reactar' was made because:
    (a) Heavy water was available ingreat quantity in India
    (b) A complete industrial & technological facility was available to made such type of reactors
    (c) The reactor with Heavy water system produces more energy
    (d) It was thorium, which is in great quantities in India.

64. India's first Test Nuclear Reactor is:
    (a) Apsara (b) Cyrus
    (c) Zerlina (d) Pwchima.

65. India's first Atomic Reactor Apsara is situated at:
    (a) Kota (b) Naraura
    (c) Trombay (d) Kalpakkam.

66. India's first 'Atomic Reactar, Apsara' be come operational in the year:
    (a) 1951 (b) 1954
    (c) 1956 (d) 1960.

67. Which one of the following was the first 'Atomic Reactor' of its type in Asia?
    (a) Apsara (b) Cyrus
    (c) Zerlina (d) Purnima.

68. Which country cooprated in developing 'cyrus' reactar?
   (a) Britain (b) America
   (c) Canada (d) Japan.

69. Which of the following 'Atomic Test Reactor' is not established in Trombay?
   (a) Cyrus (b) Zerlina
   (c) Dhruv (d) Kamini.

70. A'Test Reactor' which was closed in the year 1983,
   (a) Purnima (b) Zerlina
   (c) Dhruv (d) Kamini.

71. Which one of the following is not matching?
   (a) Apsara-Uranium (b) Cyrus-uranium
   (c) Purnima-Plutonium (d) Kamini-Plutonium.

72. The reactor based on Thorium Uranium 233 Cycle is:
   (a) Cyrus (b) Dhruv
   (c) Purnima (d) Kamini.

73. The first Atomic Electricity Station of the country was set up at:
   (a) Banglore (b) Tarapur
   (c) Rawatbhata (d) Kalpakkam.

74. For which Atomic Electricity Station, the fuel is made available by America:
   (a) Tarapur (b) Rawatbhata
   (c) Kalpakkam (d) Naraura.

75. Which fuel and controlling agent is being used in Rawatbhata Atomic Station?
   (a) Thorium-Cadmium (b) Uranium-Heavy water
   (c) Uranium-Cadmium (d) Plutonium-Heavy water.

76. Which Atomic Electricity Station is made with the co-operation of Canada?
   (a) Tarapur (b) Narora
   (c) Kalpakkam (d) Rawatbhata.

77. 'Indian Nuclear Energy Corporation' was set up in

(a) 1982 (b) 1987

(c) 1989 (d) 1991.

78. Which of the following states are being benefitted by the electric-energy produced at Tarapur Atomic Electric Station?

(a) Maharashtra & Karnataka

(b) Maharashtra & Kerala

(c) Maharashtra & Gujarat

(d) Maharashtra & Madhya Pradesh.

79. At present, India has Nuclear Energy Capacity of producing'

(a) 250 Megawatt (b) 500 Megewatt

(c) 750 Megawatt (d) 1000 Megawatt.

80. At which place, Atomic Electricity Station is situated in Karnataka?

(a) Bangalore (b) Manglore

(c) Caiga (d) Bellari.

81. Write the following names in the Chronological order of their manufacturing and programming

(a) APPLE (b) IGMD

(c) STEP (d) SITE.

(a) B,C,D,A (b) B,C,A,B

(c) C,A,D,B (d) A,C,D,B.

82. When INSAT-1D was launched?

(a) 10 June, 1990 (b) 12 June, 1990

(c) 12 June, 1990 (d) 15 June, 1990.

83. India's first Remote Sensing Satellite was launched from:

(a) Bekaunur (b) Cap Kennedy

(c) French Guinea (d) Sri Harikota.

84. Cryogenic Engine is being used in

(a) Atomic-reactor (b) Fast train' engine

(c) Boeing (d) Space shuttle.

85. Statement

(A) India's telecomm unication system is linked with satellite Cause

(R) India is preparing for stay-way.

(a) Both A&R are correct and R is the true explanation of A.

(b) Both A&R are correct, but R is not correct explanation of A.

(c) A is correct but R is wrong

(d) A is incorrect but R is correct.

86. Which satellite was launched by India in 1996?

(a) IRS-P2 (b) IRS-P3

(c) IRS-EN (d) IRS-ID.

87. Indian Space Research Organisation (ISRO) was organised in the year

(a) 1957 (b) 1962

(c) 1966 (d) 1969.

88. The Chairman of 'Indian Space Research Organisation' organised in the Year 1962, was:

(a) Dr. Shanti Swarup Bhatnagar

(b) Dr. H. J. Bhabha

(c) Dr. Vikram Sarabhai

(d) Dr. M.S. Swaminathan.

89. ISRO, Space Commission and Space Department came into existence consequently as in the years:

(a) 1963, 1965, 1965 (b) 1969, 1972, 1972

(c) 1963, 1965, 1969 (d) 1969, 1972, 1975.

90. Which one of the following is not the objective of 'Indian Space Programme'

(a) Satellite's remote sensing power for surveying resources

(b) Environmental monitoring

(c) Services related to weather science

(d) None of the above.

91. The headquarter of ISRO is situated at:
    (a) Bangalore (b) Delhi
    (c) Ahmedabad (d) Tiruavanthpuram.

92. 'ISRO' is the research wing of
    (a) Defence ministry
    (b) Agriculture ministry
    (c) Department of Meteorology
    (d) Department of Space.

93. Which one of the following is not matched correctly?
    (a) Vikram Sarabhi Space Centre, Tiruavanthpuram
    (b) Isro Satellite .Application Centre, Bangalore
    (c) Space Application Centre, Delhi
    (d) Shar Centre, Sri Harikota.

94. The main launching centre of ISRO is :
    (a) INTRAC (b) SHAR
    (c) SAC (d) ISAC.

95. ISRO Telemetry Regulation & Assessment Network is situated at:
    (a) Delhi (b) Dehradun
    (c) Bangalore (d) Sri Harikota.

96. Which one of the following is providing telemetry regulation facility to ISROS launching rockets & satellite projects?
    (a) SAC (b) ISAC
    (c) VSSC (d) INTRAC.

97. Capacity of launching satellites in the space is about
    (a) 5 (b) 6
    (c) 7 (d) 18.

98. The Rocket launching Centre was set up at Thumba in the year:
    (a) 1962 (b) 1963
    (c) 1965 (d) 1967.

99. The name of the first Satellite of India is :
    (a) Bhaskar (b) Rohini
    (c) Aryabhatta (d) Apple.

100. For running SITE programme India sought help from which of the following countries:
    (a) Japan (b) Russia
    (c) America (d) France.

101. The satellite 'Aryabhatta' was launched from
    (a) India (b) Japan
    (c) Soviet Union (d) America.

102. The Satellite 'Bhaskar-1' was designed and manufactured by:
    (a) ISRO (b) UNEP
    (c) WHO (d) NASA.

103. Rohini was of weight
    (a) 22 Kg. (b) 32 Kg.
    (c) 42 Kg. (d) 52 Kg.

104. India's multipurporse Satellite series is named as
    (a) APPLE (b) INSAT
    (c) ASLY (d) SHAR.

105. Rakesh Sharma entred in Space under
    (a) Indo-Japan joint venture
    (b) Indo- American joint venture
    (c) Indo-Russia joint venture
    (d) Purely Indian venture.

106. Insat-1B was launched by
    (a) Ariane (b) Challenger
    (c) SLV-3 (d) ASLV-D1.

107. INSAT-1B is oprational:
    (a) From 1982 to 1983 (b) From 1983 to 1990
    (c) From 1988 to 1989 (d) From 1987 to 1994.

108. Which one of the following was the first multipurpose Satellite:
(a) INSAT-1C (b) INSAT-2D
(c) INSAT-2A (d) None of these.

109. Kalpana Chawla went into the Space by which one of the following space-Shuttles:
(a) INSAT-1C (b) INSAT-1D
(c) INSAT-2A (d) None of these.

110. 'SITE' programme is related to:
(a) Meteorology (b) Defence programmes
(c) Public education (d) Telecommunication.

111. Which one of the following was the first national geo-static sattellite?
(a) Rohini (b) Aryabhatta
(c) Oceansett (d) Apple.

112. INSAT programme is being executed by
(a) Deptt. of Space
(b) Deptt. of Telecommunication
(c) Deptt. of Meteorology
(d) All of the above mentoned.

113. INSAT System is under the responsibility of
(a) Space Department
(b) Telecommunication Department
(c) Meteorology Department
(d) All of the above.

114. In the first series of INSAT which satellites have worked solidly?
(a) INSAT-1A,1B (b) INSAT-1B,1C
(c) INSAT-1B,1D (d) INSAT-1A,1D.

115. INSAT-2A is expected to work for
(a) 5 Yrs. (b) 7 Yrs.
(c) 9 Yrs. (d) 11 Yrs.

116. Which one of the following satellites made facility of telephone or STD calls?

(a) INSAT-1D (b) INSAT-2A
(c) INSAT-2B (d) INSAT-2C.

117. The direct telecasting became possible in Middle Asia, South Asia West Asia by:

(a) INSAT-1D (b) INSAT-2A
(c) INSAT-2B (d) INSAT-2C.

118. Which of the following Satellite became defunctional soon:

(a) INSAT-2A (b) INSAT-2B
(c) INSAT-2C (d) INSAT-2D.

119. Which satellite of the following crosses India four times in a day?

(a) INSAT-1D (b) INSAT-2C
(c) IRS-1C (d) IRS-1D.

120. IRS-PU satellite is also called as

(a) INSAT (b) Ocean-SAT
(c) Kit-SAT (d) Tub-SAT.

121. Advance Satellite Launching Vehicle can launch a Satellite in the low earth-circle, which is consisting the weight between

(a) 35-50Kg. (b) 100-150 Kg.
(c) 500-600 Kg. (d) 1000-1200 Kg..

122. Remote Sensing Satellite system developed by India, placed it in:

(a) Forth Country to do so
(b) Fifth Country to do so
(c) Sixth Country to do so
(d) Seventh Country to do so.

123. The first launching vehicle developed wholly by India is:

(a) ASLV-D4 (b) PSLV-02
(c) PSLV-D3 (d) PSLV-C1.

124. Which satellite has been launched by PSLV-C1 launching vehicle?
(a) IRS-P4 (b) IRS-1D
(c) INSAT-2D (d) Kit- SAT.

125. Which satellite was /were launched by PSLV-2 launching vehicle of India?
(a) Ocean-SAT (b) Kit- SAT
(c) Tub-Sat (d) All of these.

126. The Satellite IRS-PU was of:
(a) 1,200 Kg. (b) 870 Kg.
(c) 1050 Kg. (d) 922 Kg..

127. PSLV-2 launched Satellite in:
(a) 1992 (b) 1993
(c) 1994 (d) 1999.

128. The most enriched variety of radimotr is fitted in-
(a) INSAT-1B (b) INSAT-1C
(c) INSAT-1D (d) INSAT-1E.

129. The first Indian made Cryogenic engine was tested in
(a) Bangalore (1996) (b) Mahendragiri (1996)
(c) Bangalore (1998) (d) Mahendragiri (1998).

130. On the pressure of which of these following countries, Russia denied to give India the required Cryogenic engine
(a) Japan (b) France
(c) America (d) China.

131. Which of the following is the Short-Range Missile of India?
(a) Agni (b) Arjun
(c) Nag (d) Prithvi.

132. Name that Missile, which was tested jointly by Indian scientiest and Defence personnels from Chandipur on 27.1.1996:
(a) Prithvi-1 (b) Prithvi-II
(c) Akash (d) P.S.L. V, D-2.

133. The tank, made in India is
(a) Indra (b) Akash
(c) Arjun (d) Mahabli.

134. Which one of the following is the surface-to Surface Indian Defence Missile?
(a) Nag (b) Trishul
(c) Akash (d) Prithvi.

135. Which one of the following is not a missile?
(a) Agni (b) Akash
(c) Pinak (d) Nag.

136. Which one of the following is the latest tank?
(a) Bhim (b) Akash
(c) Arjun (d) Prithvi.

137. The date of succeseeful testing of medium range missile 'Agni' is:
(a) 21 April, 1998 (b) 23 June, 1989
(c) 22 May, 1989 (d) 24 December, 1989.

138. Anti tank missile 'Nag' was launched from,
(a) Chandipur (b) Midnapur
(c) Madurai (d) New Delhi.

139. On April 11, 1999, India made a successful testing of Agni-II Its striking capacity with a weight of 1000 Kg on it, is:
(a) 500 km (b) 1,000 km
(c) 2,000 km (d) More than 2000 km.

140. The scientist with major contribution in developing Agni-II is :
(a) Dr. Ashutosh Ghosh
(b) Dr. M.L. Khan
(c) Dr. A.P. J. Abdul Kalam
(d) Dr. M.S. Swaminathan.

141. From which state Agni-II was launched?
(a) Uttar Pradesh (b) Maharashtra
(c) Orissa (d) Rajasthan.

142. The striking capacity of 'Trishul' missile is:

(a) 50 km. (b) 500 km.
(c) 4,000 km. (d) 5,000 km..

143. 'Trishul' missile was tested on

(a) 6 April, 1999 (b) 16 April, 1999
(c) 16 May, 1999 (d) 16 June, 1999.

144. 'Pinak' is a

(a) A Surface-to-Air missile
(b) A Multibarrel Rocket system
(c) A submarrine
(d) A war-tank.

145. Defence Research and dovelopment organisation was organised in the year

(a) 1948 (b) 1958
(c) 1968 (d) 1978.

146. 'India' is name of :

(a) A missile
(b) A tank
(c) A pilot less aeroplane developed by Defence
(d) A Radar.

147. The pilot less aeroplane developed by Defence Research Development organisation is:

(a) Sagarika (b) Ajeya
(c) Nishant (d) Indra.

148. Integrated Guided Missile Development Programme was started in the year

(a) 1963 (b) 1973
(c) 1983 (d) 1989.

149. Integrated Missile Programme includes the no. of laboraterіes of DRDO are;

(a) 12 (b) 17
(c) 19 (d) 24.

150. The first missile manufactured wholly in the country is :

(a) Patriot (b) Scud
(c) Prithvi (d) Trishul.

151. Which of the following is not matched correctly?

(a) Trishul — Surface to Air Strike
(b) Prithvi — Surface to Surface Strike
(c) Akash — Surface to Surface Strike
(d) Agni — Surface to Air Strike

152. In which missile both solid and liquid fuel are being used?

(a) Prithvi (b) Trishul
(c) Agni (d) Nag.

153. Agni missile uses the type of fuel

(a) In both stages — liquid
(b) In both stages — solid
(c) In first stage — solid & second stage liquid
(d) In first stage — liquid & second stage Solid.

154. Agni — Tank missile's name is:

(a) Agni (b) Akash
(c) Nag (d) Prithvi.

155. Who is named as India's Missile Man?

(a) Dr. Kasturi Rangan
(b) Dr. Yashpal
(c) Dr. A. P. J. Abdul Kalam
(d) Dr. M. S. Swaminathan.

156. Which one of the following has striking capacity of more than 3000 km?

(a) Trishul (b) Nag
(c) Agni-1 (d) Agni-II.

157. A missile which can strike five aeroplanes at a time is

(a) Prithvi (b) Akash
(c) Nag (d) Agni.

158. Which missile of the following can be launched from a helicopter
   (a) Prithvi (b) Agni
   (c) Nag (d) Akash.

159. 'Nag' has the striking capacity of
   (a) 2km (b) 4km
   (c) 12km (d) 14km.

160. The largest official authority to manufacture defence equipments is
   (a) Hindustan Aeronautics Ltd.
   (b) Bharat Dynamics Ltd.
   (c) Goa Shipyard Ltd.
   (d) Defence factories organisation.

161. In which year, British government Agricultural Department in different provinces?
   (a) 1857 (b) 1863
   (c) 1871 (d) 1881.

162. Imperial Agricultural Research Institute was set up at Pusa in the year
   (a) 1901 (b) 1905
   (c) 1919 (d) 1926.

163. Indian Agricultural Research Institute is situated at:
   (a) Hyderabad (b) Calcutta
   (c) Mumbai (d) Delhi.

164. Indian Agricultural Research Institute Delhi was transferred to Delhi in the year 1936. Before that, it was situated at:
   (a) Maharashtra (b) Karnataka
   (c) West Bengal (d) Bihar.

165. Who was chairman of Royal Commission of Agriculture, which was organised in the year 1926?
   (a) Lord Wavell (b) Lord mountbatten
   (c) Lord Linlithgo (d) Lord clive.

166. Imperial Council of Agricultural Research was organised in the year

(a) 1919 (b) 1926

(c) 1929 (d) 1966.

167. When was the name 'Imperial Council of Agricultural Research' changed into 'Indian Council of Agricultural Research'

(a) 1919 (b) 1926

(c) 1929 (d) 1946.

168. When ICAR was given full autonomy?

(a) 1947 (b) 1950

(c) 1966 (d) 1973.

169. The first Director General of qutonomous ICAR was

(a) Dr. Gajendra Gadkar

(b) Dr. V. P. Pant

(c) Dr. M. S. Swaminathan

(d) Dr. Hargovind Khurana.

170. The aim of ICAR is not one among the following:

(a) To enhance agricultural research

(b) To gather informations related to Agricultural Science

(c) To publish research thes is about agricultural Science

(d) None of the above.

171. ICAR works on agricultural research and education under

(a) Education ministry

(b) Agricultural ministry

(c) Food ministry

(d) Education & Agriculture ministry.

172. The chief administrator of ICAR is

(a) Agriculture minister

(b) Agriculture State minister

(c) Director General

(d) Prime Minister.

173. The no. of Deputy Directors to assist ICAR's Director General are

(a) Two (b) Three
(c) Four (d) Five.

174. In how many regions the whole country is divided into?

(a) Five (b) Six
(c) Seven (d) Eight.

175. 'Sugarcane Cross Breeding Institution' is situated at:

(a) Coimbatore (b) Lucknow
(c) Karnal (d) Muzaffarnagar.

176. Central Rice Research Institution is situated at:

(a) Nagpur (b) Krrukshetra
(c) Katak (d) Calcutta.

177. The whole responsibility of establishing an Agriculural University lies on:

(a) ICAR (b) State government
(c) Both a&b (d) None of these.

178. After five years of establishment of Agricultural university who is responsible to run it?

(a) State government (b) District Administration
(c) District Panchayat (d) Municipal Corporation.

179. Which one of the following is not correctly matched?

(a) Indian Horticulture Research Institute- Jammu
(b) Central Potato Research Institute- Allahabad
(c) Centraln cotton Research Institute- Nagpur
(d) Indian Agriculture Research Institute- Banglore.

180. Which one of the following is situated at Lucknow?

(a) JAR (b) CRRI
(c) CTRI (d) IISR.

181. Which one of the following crops has been effected by Green Revolutoin to the highest level?

(a) Paddy (b) Maize
(c) Millet (d) Wheat.

182. Who is responsible for Green Revolution?
(a) Dr. Homi J. Bhabha
(b) Dr. Vikram Sarabhai
(c) Dr. M. S. Swaminathan
(d) Dr. C. V. Raman.

183. When Green Revolution started in India?
(a) 1955-56 (b) 1966-67
(c) 1970-71 (d) 1971-72.

184. The success of 'Green Revolution' depends on the availability of
(a) Fertilizer & Pesticide
(b) Better irrigation facility
(c) High quality seed
(d) All of these.

185. The main basis of 'Green Revolution' is
(a) Community development
(b) Removal of Zamindari system
(c) Improved Agriculture
(d) Land consolidation.

186. When 'Green revolution' was launched in India?
(a) 1960's (b) 1970's
(c) 1980's (d) 1950's.

187. India became self dependant is food production in the year?
(a) 1966-67 (b) 1970-71
(c) 1976-77 (d) 1980-81.

188. The inrease in the production of food crops in the year 1997 is
(a) Two times (b) Three times
(c) Four times (d) Five times.

189. The Indian govt. started IADP in the year:
(a) 1955 (b) 1960
(c) 1965 (d) 1970.

190. The dwarf variety of wheat received from Japan and which become the basic point of 'Green Revolution' is named as

(a) Barelo- 6 (b) Java
(c) Norin- 10 (d) William- 4.

191. The variety of wheat developed in Mexico, which is even more dwarf to Norin-10 variety is invented by-

(a) Dr. S.P. Soloman
(b) Dr. M.S. Swaminathan
(c) Dr. Norman E. Borlag
(d) Dr. Blackman.

192. Mexico achieved self-dependence in wheat production in the year

(a) 1946 (b) 1950
(c) 1956 (d) 1963.

193. In the beginning of Green Revolution, the four dwarf varieties of wheat given by Dr. Borlaug includes

(a) Larma Rosa (b) Sohara 64
(c) Mayo 64 (d) All of these.

194. The dwarf variety of wheat received from Mexico, was suffering from a defficiency of:

(a) Late riping
(b) grain of red colour
(c) not a disease-resistant variety
(d) not sensitive of fertilizer.

195. The native coloured wheat variety developed by Dr. Swaminathan from red coloured wheat is named as:

(a) Sonora 63 (b) Sonora 645
(c) Sharbati Sonora (d) Jaya.

196. Which variety of wheat is mostly sown in India now a days?

(a) Sonora-63 & 64
(b) Sonora-63, Shorbati Sonora
(c) Sharbati Sonora
(d) India.

197. The dwarf quality of rice was first developed in
(a) Japan (b) Philippines
(c) Taiwan (d) India.

198. IR= 8 paddy has been developed by
(a) Crossing De-Jio-yu-Jane variety with T.N_1
(b) Crossing 'De-Yu-Jane' variety with 'Padma'
(c) Crossing T.N.-1 variety with Padma
(d) Crossing T. N-1variety with peta.

199. The first semi-dwarf variety of paddy developed in India is resistant is known as:
(a) Kaveri (b) IRB
(c) Wala (d) Jaya.

200. Which Agricultural scientist got Nobel Prize for peace:
(a) Dr. Pohalman (b) Dr. Borlog
(c) Dr. Swaminathan (d) Dr. Paul.

201. Dr. Swaminathan was honoured with
(a) World-food Prize (b) Nobel Prize
(c) Jamnalal Bajaj Award (d) Gandhi Shanti Award.

202. Which of the following was essential factor for Green Revolution?
(a) Soil test
(b) Chemical fertilizer
(c) Loan to farmars by Banks
(d) All of these.

203. 'Green Revolution' caused
(a) Soil pollution intensification
(b) Decreased soil-pollution
(c) No change of soil's feature
(d) Soil pollution increased and natural fertilization decreased.

204. As a result of 'Green Revolution'
(a) Bio features increased
(b) Bio-features decreased

(c) No difference on bio features
(d) Cannot say.

205. It started yielding sufficient food due to 'Green Revolution' but:
(a) There is a bad effect on commerical Crop
(b) Unemployment increased
(c) Pollution increased
(d) All of the above are correct.

206. If a 'height increasing gene' helps in producing a mouse which is 8 times of a normal mouse it will be called as:
(a) Cross breading
(b) Genetic Engineering
(c) Interchangeable breeding
(d) Harmonisation.

207. Which of the following works come under Bio-technology area?
(a) Use of bios or bio-products in industrial process
(b) To modernise the process of those commercial industries, which are producing raw-materials for being used in bio-technological research.
(c) To use modern technology for research in the field of bio-irregularities .
(d) To use modern technology for enhancing biospheres.

208. Table 'A' is statement and 'R' is reason for that:

Statement (A): If scientists search out a hair of Einstein and draw out D.N.A. from that, they may produce second Einstein by cloning process.

Reason (R) : In the first stage of cloning D.N.A. is extracted from the cell of a foetus and is transferred into some special cells and these eggs are then transplanted in some mothers vagina, so that a similar child could be brought forth.

Which of the following options is correct in reference to the above two statements?

(a) Both 'A' and 'R' are correct and R is the correct explanation of A:

(b) Both 'A' and 'R' are correct but R is not the, correct explanation of A
(c) A is correct and R incorrect
(d) A is incorrect but R correct.

209. Which of the following are used as bio-fertilizer?
1. Azona
2. Blue groon Algore
3. Alpha-Alpha

(a) Only 2 (b) 1 &2
(c) 1 &3 (d) 1, 2, & 3.

210. To establish paternity, which of the following technique is used?
(a) Protein analysis
(b) Counting of chromosomes
(c) DNA quantitive analysis
(d) DNA Finger printing.

211. In the following micro-organism, which one is mostly used?
(a) Micro organs
(b) Micro organs and fungi
(c) Micro organs and Algae
(d) micro organs and microalgae and fungi.

212. When was Deptt. of Biotechnology set up in India?
(a) 1984 (b) 1986
(c) 1988 (d) 1990.

213. Department of Biotechnology is under
(a) Ministry of Biotechnology
(b) Finance Ministry
(c) Ministry of Agriculture
(d) Science & Technology ministry.

214. All dimenslonal information of Bio-technology can be received from:
(a) Bio-technological network
(b) Central Statistics Organisation

(c) Prassar Bharati
(d) Akashvaani.

215. '2' Tissue Culture Pilot plants' are set up in
(a) New Delhi & Ahmedabad
(b) New Delhi & Chennai
(c) New Delhi & Pune
(d) New Delhi & Bangalore.

216. Restrictive enzyme was developed by:
(a) Waxman
(b) Alexander Flemming
(c) Berg
(d) Smith & Nathenes.

217. Humulin is
(a) An acid (b) A protein
(c) A carbohydrate (d) A Fat.

218. Who invented couplet DNA by mixing two different DNAs?
(a) Paul Berg (b) Paul Berg
(c) Nathans (d) Nathans.

219. When was the first insulin reached in the market, which was made using Bio-technique?
(a) 1972 (b) 1977
(c) 1982 (d) 1987.

220. Which of the following is Institute of Biotechnology of India?
(a) NII (b) IARI
(c) ICAR (d) CFTRI.

221. How many Distribution Centres are there in the Biotechnology Information System Network of India?
(a) 5 (b) 10
(c) 15 (d) 20.

222. In which of the following places the national facility of Graphis & atomic modelling is available apart from BTIS Net?
   (a) Microorganic Technology Institute, Chandigarh
   (b) India Disease Resistant Institute, New Delhi
   (c) Madurai Kamraj University, Madurai
   (d) Indian Agriculture Research University, New Delhi.

223. The factory for manufacturing Polio-dose is situated at:
   (a) Raibareli (b) Bulandshahar
   (c) Meerut (d) Bijnour.

224. Where the production of Hepatitis vaccination started first in India?
   (a) Bangalore (b) Bulandshahr
   (c) Meerut (d) Bijnour.

225. Which one of the following antibiotics is not being produced in India?
   (a) Penicillin (b) Ampicillin
   (c) Chloremphenicd (d) Tetramycilline.

226. Where in India, the facility for storing blood cells at low temprature is available?
   (a) Pant Hospital, Delhi
   (b) All India, Institute of Medical science, Delhi
   (c) Indian Institute of Haemetology, Mumbai
   (d) Red Cross Society Headquarter.

227. The fully indigenous Glucometer on the basis of Biosensor technique has been made at:
   (a) IIT, Delhi
   (b) IIT, Kharagpur
   (c) Central Electro chemical Research Institute, Karaikudi
   (d) Indian Institute of Haemetology, Mumbai.

228. Micro Organic Technological Institute & Delhi University has developed
   (a) Medicinal banana (b) Medicinal tomato
   (c) Medicinal papaya (d) All of these.

229. Which laboratory of India has got success in blooming of flower in a Bamboo tree at earlier?
   (a) Lupin Laboratory
   (b) National Chemical Laboratory, Pune
   (c) Tata Energy Research Institute
   (d) A.B. Thomas & Co..

230. When Indian Government started National Biotechnology Board
   (a) 1980 (b) 1981
   (c) 1982 (d) 1984.

ANSWER SHEET

1.	(a)	2.	(c)	3.	(c)	4.	(c)
5.	(a)	6.	(d)	7.	(a)	8.	(d)
9.	(c)	10.	(b)	11.	(b)	12.	(c)
13.	(a)	14.	(d)	15.	(b)	16.	(c)
17.	(c)	18.	(c)	19.	(d)	20.	(c)
21.	(c)	22.	(b)	23.	(d)	24.	(b)
25.	(d)	26.	(b)	27.	(d)	28.	(d)
29.	(a)	30.	(c)	31.	(a)	32.	(b)
33.	(b)	34.	(c)	35.	(a)	36.	(b)
37.	(c)	38.	(a)	39.	(b)	40.	(d)
41.	(d)	42.	(c)	43.	(c)	44.	(c)
45.	(c)	46.	(c)	47.	(d)	48.	(a)
49.	(c)	50.	(c)	51.	(d)	52.	(d)
53.	(a)	54.	(a)	55.	(d)	56.	(b)
57.	(c)	58.	(d)	59.	(d)	60.	(b)
61.	(a)	62.	(b)	63.	(b)	64.	(a)
65.	(c)	66.	(c)	67.	(a)	68.	(c)
69.	(d)	70.	(b)	71.	(d)	72.	(d)
73.	(b)	74.	(a)	75.	(b)	76.	(d)
77.	(b)	78.	(c)	79.	(c)	80.	(c)
81.	(b)	82.	(b)	83.	(a)	84.	(d)
85.	(c)	86.	(b)	87.	(b)	88.	(c)
89.	(b)	90.	(d)	91.	(a)	92.	(d)
93.	(c)	94.	(b)	95.	(c)	96.	(d)
97.	(c)	98.	(b)	99.	(c)	100.	(c)

101.	(c)	102.	(a)	103.	(b)	104.	(b)
105.	(c)	106.	(b)	107.	(b)	108.	(d)
109.	(d)	110.	(c)	111.	(d)	112.	(d)
113.	(a)	114.	(c)	115.	(c)	116.	(c)
117.	(d)	118.	(d)	119.	(c)	120.	(b)
121.	(b)	122.	(b)	123.	(b)	124.	(b)
125.	(d)	126.	(c)	127.	(d)	128.	(d)
129.	(d)	130.	(c)	131.	(c)	132.	(b)
133.	(c)	134.	(d)	135.	(c)	136.	(c)
137.	(c)	138.	(a)	139.	(d)	140.	(c)
141.	(c)	142.	(c)	143.	(b)	144.	(b)
145.	(b)	146.	(d)	147.	(c)	148.	(c)
149.	(b)	150.	(c)	151.	(c)	152.	(c)
153.	(c)	154.	(c)	155.	(c)	156.	(d)
157.	(b)	158.	(c)	159.	(b)	160.	(d)
161.	(d)	162.	(b)	163.	(d)	164.	(d)
165.	(c)	166.	(c)	167.	(d)	168.	(c)
169.	(b)	170.	(d)	171.	(b)	172.	(c)
173.	(c)	174.	(d)	175.	(a)	176.	(c)
177.	(c)	178.	(a)	179.	(c)	180.	(d)
181.	(d)	182.	(c)	183.	(d)	184.	(d)
185.	(c)	186.	(a)	187.	(b)	188.	(c)
189.	(b)	190.	(c)	191.	(c)	192.	(c)
193.	(d)	194.	(b)	195.	(c)	196.	(d)
197.	(c)	198.	(d)	199.	(d)	200.	(b)
201.	(a)	202.	(d)	203.	(d)	204.	(a)
205.	(d)	206.	(a)	207.	(a)	208.	(a)
209.	(a)	210.	(a)	211.	(a)	212.	(a)
213.	(a)	214.	(a)	215.	(a)	216.	(a)
217.	(a)	218.	(a)	219.	(a)	220.	(a)
221.	(a)	222.	(a)	223.	(a)	224.	(a)
225.	(a)	226.	(a)	227.	(a)	228.	(a)
229.	(a)	230.	(a)				

SET I
ASSESS YOURSELF

1. Zero degree Kelvin or absolute zero is equivalent to
   (a) 142°C (b) 93°C
   (c) -273°C (d) -97°C.

2. The S.I. unit for measuring potential difference is
   (a) Ohm (b) Volt
   (c) Watt (d) Candela.

3. Which compound is used as a flux in soldering?
   (a) Copper Sulphate
   (b) Calcium Nitrate
   (c) Zinc Chloride
   (d) Potassium Sulphate.

4. The Science of production of desired off springs is called
   (a) Eugenics (b) Embroylogy
   (c) Ethnalogy (d) Ethnography.

5. The first man to visit south pole was?
   (a) Robert Piery (b) Thomas Pike
   (c) Ralphinston (d) Amundsen.

6. The first satellite of India 'Aryabhatt' was launched from USSR in
   (a) 1975 (b) 1976
   (c) 1977 (d) 1978.

7. Who can be called the father of modern computers?
   (a) Blaise Paskal (b) Heibneiz
   (c) Calmer (d) Charles Babbage.

8. Who discovered super conductivity?
   (a) Meisner (b) Dchsenfeld
   (c) Onnes (d) Muller.

9. The discoverer of Asprin was
   (a) Banting (b) Dresser
   (c) Watson (d) Draker.

10. Which among the following gas contributs the least in Green house effect?
   (a) CO_2 (b) CFC
   (c) Methane (d) Nitrous Oxide.

11. Sun spats upper dark because they are
   (a) Cooler (b) Warmer
   (c) Gases (d) Liquids.

12. The size of moon is what percentage of the earth
   (a) $\frac{1}{3}$ (b) $\frac{1}{4}$
   (c) $\frac{1}{5}$ (d) $\frac{1}{6}$.

13. Water has the maximum density at
   (a) 4°k (b) 4°C
   (c) - 4°C (d) -40°C.

14. Which of the following is a semi-conductor
   (a) Chromium (b) Germanium
   (c) Copper (d) Phosphorous.

15. The Indian Institute of Science is Located in
   (a) New Delhi (b) Bangalore
   (c) Calcutta (d) Pune.

16. Who propounded the universal law of gravitaion:
    (a) Einstein (b) Newton
    (c) Galileo (d) Copernicus.

17. A balanced diet is -
    (a) Nutritionally inadequate
    (b) Nutritionally adequate
    (c) Sufficient to satisfy
    (d) Rich in fats so as to produce adequate energy.

18. A man who climbs a mountain suffers from-
    (a) Oxygen dificiency (b) Metabolic faliure
    (c) Haemophilia (d) Nitrogen dificiency.

19. The Xylam is found in-
    (a) The root (b) The stem
    (c) The heat (d) None of these.

20. Find out the name of the scientist given in the picture below
    (a) Einstein (b) Marie Curie
    (c) Maria Gaeppert Mayar (d) Gertaude Elion.

21. Agar-Agar is derived from
    (a) Algae (b) Fungi
    (c) Kerus (d) Lichen.

22. The enzyme which coagulaters the milk is
    (a) Pepsin (b) Renin
    (c) Trypsin (d) Lactose.

23. Solid Carbon dioxide is known as
    (a) Slaked ice (b) Wet ice
    (c) Dry ice (d) Carbon ice.

24. Which of the following polymer is a non-softening plastic?
    (a) Teflon (b) Orion
    (c) Bakellite (d) Teryline.

25. The neck is absent in the frog. This helps the frog in-
    (a) Respiration (b) Jumping on the ground
    (c) Swimming (d) None of these.

26. Number of eyes in an earth worm is
    (a) One (b) Two
    (c) Many (d) No eyes.

27. The end-product of the digestion to search in the alimentary canal is
    (a) Glucose
    (b) Galactose
    (c) Maltose
    (d) Ismaltose.

28. The numer of heart-beats on an average in an adult human is in the range of
    (a) 60-65 (b) 66-70
    (c) 71-80 (d) 85-90.

29. The total number of bones in man is
    (a) 212 (b) 206
    (c) 202 (d) 200.

30. The device which converts sound energy into electrical energy is called
    (a) Amplifier (b) Speaker
    (c) Microphone (d) Transmeter.

31. Electric current in a mental core is due to flow of
    (a) Electrones (b) Protons
    (c) Ions (d) Heals.

32. The material used in the fabrication of a Transistor is
    (a) Aluminium (b) Copper
    (c) Silicon (d) Silver.

33. Electricity is transmitted over long distances by high voltage AC because
    (a) It is fast
    (b) There is less wastage of energy
    (c) It is safer
    (d) It is coper.

34. When hot-water is sprinkled on a hoter glass tumbler, it breaks because
    (a) Glass suddenly expands
    (b) Glass suddenly contracts
    (c) Water evaporates
    (d) Glass reacts chemically with water.

35. Which one of the following gases has the higest calorific value?
    (a) Natural gas (b) Water gas
    (c) Coal gas (d) Indane gas.

36. The coating of a thin layer of zinc on steel or iron objects is known as
    (a) Hot dipping (b) Tinning
    (c) Galvanising (d) Electroplating.

37. Which one of the following organic compound is used to remove stains of Iron compounds?
    (a) Oxalic acid (b Bongoic acid
    (c) Phthalic acid (d) Cinnamic acid.

38. Which of the following statement is not correct? Vulcanisation of rubber makes it
    (a) Soft and flexible
    (b) Strong and hard
    (c) Chemical resistant
    (d) Withstand higher temperatures.

39. Which one of the following is not a fertiliser?
    (a) Ammonium Sulphate
    (b) Calcium Sulphate
    (c) Calcium Ammonium Sulphate
    (d) Potassium nitrate.

40. Dialysis is used to perform the function of-
    (a) Lungs (b) Liver
    (c) Heart (d) Kidneys.

41. Which of the following does not provide any energy?

(a) Fats (b) Proteins
(c) Vitamins (d) Carbohydrates.

42. The brain of a computer is

(a) CPU (b) CD
(c) Floppy disc (d) Megabyte.

43. Which of the following is not a part of the Darwin's theory of evolution?

(a) Natural Selection
(b) Struggle for existence
(c) Survival of the fittest
(d) Inheritance of aquired characters.

44. The nitrogen in the ecosystem is circulated by

(a) Earth worms (b) Bacteria
(c) Protozoa (d) Fungi.

45. The presence of air cavities is an adaptation of

(a) Desert plants (b) Trees
(c) Water plants (d Mesophytes.

46. Acupuncture is

(a) A disease of heart
(b) Servicing of tubes and Tyres
(c) A treatment method with needles
(d) A crope culture.

47. Air coolers are more suitable for

(a) Hot and humid climate
(b) Hot and dry climate
(c) Cool & humid climate
(d) Cool & dry climate

48. Amoebiasis is causing

(a) Dysentery (b) Fever
(c) Severe cold (d) Headache and cold.

49. Some viruses have R.N.A. but no D.N.A. This would indicate that
    (a) These viruses connot replicate
    (b) These viruses have no heritable infarmation
    (c) R.N.A. transmits the hereditary information in these viruses
    (d) Their nucleic acids can be crystallised.

50. Slag is a name given for
    (a) Malten Iran
    (b) Malten sand
    (c) Malten Alumina
    (d) Malten Calcium silicate.

ANSWER SHEET

1.	(c)	2.	(b)	3.	(c)	4.	(a)
5.	(d)	6.	(a)	7.	(d)	8.	(c)
9.	(b)	10.	(d)	11.	(a)	12.	(b)
13.	(b)	14.	(b)	15.	(b)	16.	(d)
17.	(b)	18.	(c)	19.	(a)	20.	(b)
21.	(a)	22.	(b)	23.	(c)	24.	(c)
25.	(b)	26.	(c)	27.	(a)	28.	(c)
29.	(b)	30.	(c)	31.	(a)	32.	(c)
33.	(b)	34.	(a)	35.	(d)	36.	(c)
37.	(a)	38.	(a)	39.	(b)	40.	(d)
41.	(c)	42.	(a)	43.	(d)	44.	(b)
45.	(c)	46.	(c)	47.	(b)	48.	(a)
49.	(b)	50.	(d)				

SET II
ASSESS YOURSELF

1. Mercury is used in therometers because
   (a) It is available in pure form
   (b) It does not stick to the glass walls.
   (c) Its expansion is uniform and accurate
   (d) All of the above.
2. Temporary hardness of water is due to the presence of
   (a) Sulphates of Calcium and Sodium
   (b) Chlorides of Calium and Magnesium
   (c) Hydroxides of Calcium and Sodium
   (d) Bicarbonates of Calcium and Magnesium.
3. It is always seen that at the source of electric generation, electricity is supplied at high voltage why?
   (a) No one can use electricity at high voltage by cheating
   (b) There is less power loss at the source of electricity generation
   (c) Both of the above
   (d) None of these.
4. Which of the following is incorrectly matched?
   (a) Blood circulation - William Harvey
   (b) Vaccination- Louis Pasteur
   (c) Penicillin- Alexander Fleming
   (d) Theory of Evolution- Charles Darwin.

5. Which of the following is a function of carburettor?
   (a) It controls the fuel.
   (b) It keeps the engine cool.
   (c) It converts alternating current to low voltage
   (d) It mixes air with petrol to give cool vapour.

6. Cream and butter can be separated from milk by adding
   (a) Alcohol (b) Water
   (c) Alkalis (d) Acid.

7. Carbon dating is used to determine the age of
   (a) Crocks
   (b) Fossils
   (c) Plants
   (d) All of the above.

8. Naphthalene is most soluble in
   (a) Water (b) Cl_2
   (c) H_2SO_4 (d) Hydrocarbons.

9. How many litres of blood is present on an average in an adult human being?
   (a) 0.5-1.0 (b) 1.0-2.0
   (c) 2.0-4.0 (d) 5.0-6.0.

10. Bessemer converter is used in making
    (a) Iron (b) Steel
    (c) Carbon (d) Magnesium.

11. Rainbow is formed due to
    (a) diffraction
    (b) reflection and refraction
    (c) dispersion and refraction
    (d) reflection, refraction and dispersion.

12. Dark blue suit seen in a candle light seems to be
    (a) brown (b) black
    (c) blue (d) green.

13. Choose the correct descending order of the following with respect to their capacity to provide calories contents per 100 gm.
    (I) Potato
    (II) Rice
    (III) Sugar
    (IV) Banana.
    (a) III, II, IV, I (b) IV, I, II, III
    (c) I, II, III, IV (d) II, III, IV, I.

14. Which of the following pairs is incoorrectly matched?
    (a) Colitis-Colon
    (b) Encephalitis-Brain
    (c) Hepatitis-Liver
    (d) Jaundice-Throat.

15. If a magentic bar is broken into two pieces, then
    (a) The part with north pole at its one end will develop south pole at the other end vice versa
    (b) Both the pieces can be joined into a single magnet
    (c) Both the parts will be demagnetised.
    (d) Both the pieces are of on use.

16. Which chemical weathering process affects limestone the most?
    (a) Sulphonations (b) Oxidation
    (c) Hydration (d) None of the above.

17. Photosynthesis in plants takes place in the presence of
    (a) Carbon monoxide (b) Sunlight
    (c) Nitrogen (d) Rain.

18. Genetic information is encoded in
    (a) Bone (b) Blood
    (c) Brain (d) DNA.

19. Which of the following is a group of metals only?
    (a) Pb, Sb, Os (b) Br, Os, P
    (c) P, Br, Pb (d) Br, S, Pb.

20. The percentage of carbon in cast iron is
   (a) 5.4- 6.0 (b) 2- 4.5
   (c) 16.0-8.5 (d) 8.5-11.0.

21. Semi- conductors form a device of
   (a) Transformers (b) Transistors
   (c) Fans (d) Bulbs.

22. Watches made on the principle of oscillating spring, when carried to the moon, will
   (a) Slow down (b) Remain unaffected
   (c) Move faster (d) Stop warking.

23. Which of the following is a good conductor of heat and electricity?
   (a) Anthracite (b) Diamond
   (c) Gravite (d) Graphite.

24. Arrange the following items in correct order of evolution:
   I. Lizard II. Frog
   III. Shark IV. Whale
   (a) I, II, IV, III (b) I, II, III, IV
   (c) III, II, I, IV (d) IV, III, II, I.

25. What is the mineral composition of diamond?
   (a) Nickel (b) Nitrogen
   (c) Carbon (d) Zinc.

26. Uranium used in nuclear reactors is
   (a) U-236 (b) U-235
   (c) U-232 (d) U-238.

27. Match the following:
   I. Lemons (A) Alginic acid
   II. Seaweeds (b) Acetic acid
   III. Grapes (c) Tartaric acid
   IV. Vinegar (d) Citric acid.
   (a) I-D, II-C, III-A, IV-B
   (b) I-D, II-A, III-C, IV-B

(c) I-C, II-D, III-B, IV-A
(d) I-B, II-C, III-D, IV-A

28. Which of the folllowing drugs is known as 'Magic Drug' and is also misused sometimes?
(a) Penicillin (b) Corticosteroids
(c) Sulphate drug (d) Calmpose.

29. Assertion (A): Magnetic compass is not reliable to find direction

Reason (R): Earth's field deeps on changing with time.
(a) Both A and R are true and R is the correct explanation of A.
(b) A is correct and R is incorrect
(c) A and R both are incorrect
(d) Both A and R are true but R is not correct explanation of A.

30. The region of a magnet towards which the lines of induction diverge, is called
(a) East pole (b) West pole
(c) South pole (d) North pole.

31. Which of the following has have maximum protein?
(a) Wheat (b) Green vegetables
(c) Soybean (d) Coconut.

32. The fastest growing plant is
(a) Eucalyptus (b) Palm
(c) Banana (d) Coconut.

33. Goitre is caused by
(a) Unhygienic and contaminated food
(b) Deficiency of iodine
(c) Heredity
(d) Stored water.

34. The medicine for which of the following diseases was invented first of all?
(a) Cholera (b) Typhoid
(c) Tuberculosis (d) Goitre.

35. Which of the following diseases is caused by unhygienic environment and contaminated food?

(a) Cancer (b) Tuberculosis
(c) Cholera (d) Goitre.

36. Which plant was chosen by Gregor John Mendel for his experiments?

(a) Pea (b) Wheat
(c) Potato (d) Gram.

37. Radius of WBC in the blood is

(a) 0.1 Mm. (b) 0.01Mm.
(c) 0.001M.m (d) 0.0001Mm..

38. Function of haemoglobin in the blood is

(a) To transport food
(b) To transport oxygen
(c) To give red colour
(d) To help in clotting.

39. Which element is found in bones?

(a) Phosphorus (b) Calcium
(c) Sodium (d) Magnesium.

40. A food is called protective if it contains

(a) Vitamins
(b) Proteins and carbohydrate
(c) Minerals
(d) Fats.

41. Night blindness is caused due to the deficiency of

(a) Vitamin A (b) Vitamin B
(c) Vitamin C (d) Vitamin D.

42. Match of the following

I. Thermograph	(A) Electric Current
II. Anemometer	(B) Atmospheric Pressure
III. Ammeter	(C) Velocity of wind
IV. Barometer	(D) Temperature variations during a period of time on a graph.

43. Which of the following elements is used for electric lamp filamants?

(a) Nickel (b) Gold
(c) Copper (d) Tungsten.

44. Incubation period is

(a) The period between infection and the appearance of a sigh of a disease.
(b) Recover period of a disease
(c) Menstrual period
(d) None of these.

45. The normal blood pressure of a human being is

(a) 120 to 80 mm
(b) 140 to 90 mm
(c) 60 to 160 mm
(d) 140 to 70 mm.

46. The number of teeth in an adult is

(a) 28 (b) 30
(c) 32 (d) 35.

47. The number of bones in human body is

(a) 202 (b) 206
(c) 210 (d) 215.

48. Echo is produced by the of sound waves

(a) Refraction (b) Absorption
(c) Reflection (d) Diffraction.

49. Coating on iron with Zinc is known as

(a) Ionisation (b) Galvanisation
(c) Electroplating (d) Vulcanisation

50. Which of the following helps in the clotting of blood?

(a) Vitamin B_2 (b). Vitamin B_1
(c) Vitamin D (d) Vitamin K

ANSWER SHEET

1.	(d)	2.	(d)	3.	(b)	4.	(b)
5.	(d)	6.	(d)	7.	(b)	8.	(d)
9.	(d)	10.	(b)	11.	(d)	12.	(b)
13.	(a)	14.	(d)	15.	(a)	16.	(c)
17.	(b)	18.	(d)	19.	(a)	20.	(c)
21.	(b)	22.	(a)	23.	(d)	24.	(c)
25.	(c)	26.	(b)	27.	(b)	28.	(d)
29.	(a)	30.	(d)	31.	(c)	32.	(a)
33.	(b)	34.	(a)	35.	(b)	36.	(a)
37.	(c)	38.	(b)	39.	(b)	40.	(b)
41.	(a)	42.	(d)	43.	(d)	44.	(a)
45.	(a)	46.	(c)	47.	(b)	48.	(c)
49.	(b)	50.	(d)				

SET III
ASSESS YOURSELF

1. Which of the following contains a high percentage ascorbic acid?
   (a) Mango (b) Orange
   (c) Tomato (d) Palak.
2. Hydrography is related to
   (a) Solids (b) Gases
   (c) Water (d) None of these.
3. A vechile moving round a circle experiences
   (a) Centrifugal Force (b) Centripetal Force
   (c) Gravitational Force (d) None of these.
4. Insulin is a patent medicine for
   (a) Diabetes (b) Tuberculosis
   (c) Malaria (d) None of these.
5. Videotapes, used to record motion picture, work on the priniple of:
   (a) Photographic Films
   (b) Magnetic tapes
   (c) Octo-electron recording
   (d) None of these.
6. Pressure cooker is suited well to
   (a) High Pressure and high temperature regions
   (b) Coastal regions

(c) Low pressure regions
(d) High pressure regions.

7. One kg of sugar is taken to Srilanka, Bangladesh and Greenland, and weighed by a two pen common balance. At which place will it weigh the heaviest?
(a) Srilanka
(b) Bangladesh
(c) Greenland
(d) It will weigh the same at all the places.

8. Match the following

I	Pedagogy	(A)	Plants
II	Pnthisilogy	(B)	Teaching
III	Numismation	(C)	Tuberculosis
IV	Phytogeny	(D)	Coins and metals.

(a) I-D, II-A, III-B, IV-C
(b) I-C, II-D, III-A, IV-B
(c) I-B, II-C, III-D, IV-A
(d) I-A, II-B, IIII-C, IV-D.

9. Assertion (A): Amino acids are bricks of proteins
Reason (R): Amino acids are derived by hydrolysis of proteins
(a) A is true, but R is not the correct explanation of A
(b) A is true, and R is the correct explanation of A
(c) Only R is true
(d) Only A is true.

10. The tree-lands in a city help in
(a) Utilising industrial waste
(b) Replenishing the atmospheric oxygen
(c) Supply of timber
(d) None of these.

11. Which of the following instruments is used to measure blood pressure?
(a) Galvanometer (b) Manometer
(c) Sphygmomanometer (d) Ammeter

12. Hydrometer is used for determination of
 (a) Relative humidity (b) Specific gravity
 (c) Sound under water (d) None of these

13. The human heart beats times in a minute
 (a) 60-70 (b) 70-80
 (c) 80-90 (d) 100-120.

14. Which of the following pairs is incorrect?
 (a) Cirrbosis — Liver
 (b) Brain fever — Nervous system.
 (c) Trachoma — Teeth
 (d) Gluchoma — Eye.

15. The following metals have been arranged in the descending order of their densities:

 Hg, Pb, Au, and Al.

 Which of the following orders is correct?
 (a) Hg > Pb > Au > Al
 (b) Au > Hg > Pb > Al
 (c) Pb > Hg > Au > Al
 (d) Al > Au > Hg > Pb.

16. Which of the following instruments offer the least resistance to the path of electric current?
 (a) Galvanometer (b) Ammeter
 (c) Voltameter (d) Electrometer.

17. Which of the following is most abundantly found in the herbage vegetables?
 (a) Proteins (b) Starch
 (c) Carbohydrates (d) Vitamins.

18. The most difficult situation in human life is caused by
 (a) Excessive temperature
 (b) Very low temperature
 (c) Very heavy rain
 (d) Very little rain.

19. The plants, which can grow on the almost completely dryland are called

(a) Thallophytes (b) Hyellophytes
(c) Hydrophonics (d) Xerophytes.

20. Match the following:

I. Newton	Crystal Dynamics
II. Einstein	Flotation Principle
III. Archimedes	Law of Gravitation
IV. C.V. Raman	Theory of Relativity

(a) I-C , II-D, III-B, IV-A
(b) I-A, II-B, III-C, IV-D
(c) I-B, II-C, III-B, IV-A
(d) I-D, II-A, III-B, IV-C.

21. Richter Scale is used to measure

(a) Humidity
(b) Magnitude of earth quakes
(c) Specific gravity
(d) Speed of ships.

22. The rivet is used to

(a) The together two different things
(b) To smooth the surface
(c) To revolve one object about another
(d) None of these.

23. Which of the following is a semi-conductor?

(a) Copper (b Germanium
(c) Mercury (d) Silver.

24. Which of following diseases is not spread by the mosquitoes?

(a Malaria (b) Plague
(c) Yellow fever (d) Filaria.

25 Which gas is used to purify the water of ponds incities

(a) Sulphur dioxide (b) Chlorine
(c) Nitrogen (d) Ammonia.

26. The frequenty of the notes from a Flute can be increased by
    (a) Increasing the length of the pipe
    (b) decreasing the length of the pipe and decreasing the cross-section
    (c) decreasing the length of the pipe and increasing the cross-section
    (d) increasing the length and decreasing the cross-section.

27. N_2 and P_2O_5 are found in equal quantities in
    (a) DAP
    (b) ammonium phosphate
    (c) ammonium nitrophosphate
    (d) mono ammonium phosphate.

28. Match the following:

I	Alexender Flemming	(a)	Dynamite
II	Graham Bell	(b)	Telephone
III	T.A. Edison	(c)	Bulb
IV	A. Nobel	(d)	Pencillin.

    (a) I-A, II-C, III-B, IV-D,
    (b) I-C, II-B, III-D, IV-A
    (c) I-D, II-B, III-C, IV-A
    (d) I-D, II-C, III-B, IV-A.

29. The atoms, which have different atomic weights but same atomic numbers are called
    (a) Isotopos (b) Isobars
    (c) Isotones (d) None of these.

30. Which part of the body takes part in blood circulation?
    (I) Heart
    (II) Arteries
    (III) Veins
    (IV) Lungs
    (a) I and II only (b) I and III only.
    (c) I, II and III (D) I, II, III, IV

31. The terminal colours of the rainbow are
    (a) Yellow and Red (b) Violeat and Red
    (c) Violet and Indigo (d) Indigo and Orange.

32. Intesity of earthquakes is measured with a
    (a) Barometer (b) Hydrometer
    (c) Polygraph (d) Seismograph.

33. Which of the following diseases is caused by Fungus?
    (a) Hookworm (b) Malaria
    (c) Typhoid (d) Ringworm.

34. Which of the following is a Fish
    (a) Silver Fish (b) Cuttle Fish
    (c) Shark Fish (d) Star Fish.

35. Aspirin is the common name of
    (a) Salicylate (b) Salicylic acid
    (c) Acetyl salicylic acid (d) Methyl salicylate.

36. Centigrade and Fahrenhite scales give same reading at
    (a) $-40°$ (b) $-32°$
    (c) $-273°$ (d) $100°$.

37. Which of the following pairs is Odd?
    (a) Bat and Pigeon (b) Man and Bat
    (c) Goat and Sheep (d) Cat and Dog.

38. Bronze is an alloy?
    (a) Copper and iron (b) Nickel and tin
    (c) Copper and tin (d) Aluminium and tin.

39. The Ocean appears to the blue because of the sunlight falling on it is
    (a) Scattered (b) Reflected
    (c) Absorbed (d) Refracted.

40. Which of the following is the weakest acid?
    (a) HF (b) HBr
    (c) HCl (d) HI.

41. The name of N_2O is
    (a) Acidic (b) Basic
    (c) Amphoteric (d) Neutral oxide.

42. Which of the pairs of engymes does act on proteins during digestion
    (a) Pepsin and Lipose (b) Trypsin and Ptyalin
    (c) Frypsin and Trypsin (d) Trypsin and Ptyalion.

43. A needle smeared with vaseline lowezed, slowly on the surface of water in a horizontal position, appears to float. The phenomenon is an example of
    (a) Osmosis (b) Archimedes principle
    (c) Capillarity (d) Surface tension.

44. Rate of evaporation does not depend upon
    (a) Temperature of the liquid
    (b) Surface area of the liquid
    (c) Total mass of the liquid
    (d) Surface winds over the liquid.

45. Which of the following activities is involved in the use of wedge?
    (a) Cuting paper with a pair of scissors
    (b) Opening a tin lid with the help of a spoon
    (c) Cutting of a potato into two pieced with a knife
    (d) Drawing out of water from a well using a pulley.

46. A freshly-cut potato chip would shrinks when kept for about two hours in
    (a) Salt water (b) Cold water
    (c) Warm water (d) Distilled water.

47. Which of the following illustrations is not based on the principle of pressure
    (a) Swelling of rabins when kept in water
    (b) Filling up of lotion in egy droppes
    (c) Bursting of a cycle tyre in summer
    (d) Rise of water in a drinking straw.

48. "Drinking and Driving not mix". This is because alcohol causes a/ an

(a) Decrease in speed

(b) Increase in reaction time

(c) Decrease in reaction time

(d) Increase in vision.

49. The member of two species, which are associated together for mutual benefit, reputed a case of

(a) Saprohytaism (b) Parasitism

(c) Symboisis (d) Food chain.

50. RNA synthesis

(a) Vitamins (b) Proteins

(c) Minerals (d) None of these.

ANSWER SHEET

1.	(b)	2.	(c)	3.	(b)	4.	(a)
5.	(b)	6.	(c)	7.	(d)	8.	(c)
9.	(b)	10.	(b)	11.	(c)	12.	(b)
13.	(b)	14.	(c)	15.	(b)	16.	(b)
17.	(d)	18.	(a)	19.	(d)	20.	(a)
21.	(b)	22.	(a)	23.	(b)	24.	(b)
25.	(b)	26.	(b)	27.	(c)	28.	(c)
29.	(a)	30.	(c)	31.	(b)	32.	(d)
33.	(d)	34.	(c)	35.	(c)	36.	(a)
37.	(a)	38.	(c)	39.	(a)	40.	(a)
41.	(d)	42.	(a)	43.	(d)	44.	(c)
45.	(c)	46.	(a)	47.	(a)	48.	(b)
49.	(c)	50.	(b)				

SET IV
ASSESS YOURSELF

1. Einstein was awarded Nobel Prize for his work on
   (a) General theory of Relativity.
   (b) Photoelectric effect
   (c) Special theory of Relativity
   (d) Photoelectricity.

2. Hard X-rays means X-rays having
   I. high frequency
   II. large wavelength
   III. high intensity
   which one of the above combination of statements is true?
   (a) II only
   (b) I only
   (c) II and III only
   (d) I and III.

3. Which of the following is a mixture?
   (a) Sugar (b) Water
   (c) Table salt (d) Air

4. Which of the following is a compound?
   (a) Oxygen (b) Nitrogen
   (c) Air (d) Water.

5. The path of Halley's Comet around the sun is
   (a) Hyperbolic (b) Circular
   (c) Elliptical (d) Parabolic.

6. Which one of the following quantities is constant for a body moving in a circular as orbit?
   (a) Energy (b) Acceleration
   (c) Velocity (d) Momentum.

7. Which of the following processes is responsible for the production of star energy?
   (a) Fusion (b) Fission
   (c) Artificial radioactivity (d) Chemical decomposition.

8. Which one of the following metals is most reactive with water?
   (a) Aluminium (b) Sodium
   (c) Copper (d) Manganese.

9. Wollen clothes protect the body from cold because-
   (a) Clothes is a poor conductor of heat
   (b) Clothes is a good conductor of heat
   (c) External heat rays enter the body through the woollen clothes
   (d) None of these.

10. Relative humidity of the atmosphere is measured by
    (a) Lactometer
    (b) Potentiometer
    (c) Hygrometer
    (d) Hydrometer.

11. The boiling range of kerosene oil is-
    (a) 150° - 300°c (b) 300°-450°C
    (c) 200°-400°C (d) 100°-200°C.

12. Black Holes were predicted by
    (a) Copernicus (b) Einstein
    (c) Hyermann Bondi (d) Ruther ford.

13. AC. can not be used in a/an
    (a) Amplifier (b) Transformer
    (c) Voltmeter (d) Galvanometer.

14. Which one of the following is acidic in nature?
(a) Water (b) Vinegar
(c) Lime (d) Sugar.

15. The nutrient required for luxuriant vegetation growth is-
(a) Potash
(b) Single super phosphate
(c) Urea
(d) Triple super phosphate

16. Atmospheric pressure is measured by-
(a) Hygrometer (b) Hydrometer
(c) Barometer (d) Manometer.

17. Urea is stored in the-
(a) Spleen (b) Kidneys
(c) Pancreas (d) Liver.

18. Impure blood from any parts of the body enters in to the-
(a) Left auricle (b) Right ventricle
(c) Left ventricle (d) Right auricle.

19. Ammonia is used in-
(a) Paint and varnish industry
(b) Explosives
(c) Electroplating process
(d) None of these.

20. The time period of a pendulum on moon-
(a) Is zero (b) Increases
(c) Decreases (d) Remains the same.

21. Which waves can not be transmitted through vacuum?
(a) Light (b) Heat
(c) Electromagnetic (d) Sound.

22. Souring of milk is caused by
(a) Fungi (b) Yeast
(c) Microbes (d) Algae.

23. Reinforced concrete contains
    (a) Copper bars (b) Steel bars
    (c) Brass bars (d) Aluminium.

24. Oxidation of organic cells within cells accompanied by release of energy is called
    (a) Photosynthesis (b) Expiration
    (c) Respiration (d) Inspiration.

25. Ornithology is a term for study of
    (a) Bats (b) Insects
    (c) birds (d) Wind.

26. Blood sugar is the amount of .... in the circulating blood-
    (a) Sucrose (b) Lactose
    (c) Glucose (d) Galactose.

27. Which of the following is associated with the Bhopal gas tragedy?
    (a) Methyl isocyante (b) Oleum
    (c) Ethane (d) Methane.

28. Which of the following is not a primary colour
    (a) Red (b) Blue
    (c) Green (d) Black.

29. Tiredness is felt on the top of the mountains because of
    (a) High temperature
    (b) Low pressure out side the body
    (c) High pressure out side the body
    (d) Low temperature.

30. Rectifiers are used in circuits to-
    (a) Transform an alternating current into direct current
    (b) Reduce the voltage
    (c) Step up the voltage
    (d) Rectify undue current.

31. Stainless steel contains
    (a) Nickel, iron and carbon

(b) Chromium, nickel, iron and carbon
(c) Iron and carbon only
(d) Chromium, iron and carbon.

32. Which of the following is used in the baking power?
(a) Calcium carbonate (b) Sodium bicarbonate
(c) Calcium chloride (d) Sodium chloride.

33. Fluorisis is
(a) An eye disease
(b) Mottling of teeth
(c) The botanical name for a branch of flowers
(d) The process of cell division.

34. Heavy water-
(a) Is used as a coolant in the nuclear reactors
(b) Lathers with soap
(c) Is formed by prolonged electrolysis of ordinary water
(d) Is used for domestic purposes.

35. Weight of the body at the centre of the earth would be-
(a) Zero
(b) One-sixth of the weight of the body
(c) In finite
(d) One-tenth of the weight of the body.

36. Nobel prize was awarded to Albert Einstein for the discovery/ invention of the-
(a) Law of Gavitation (b) Theory of Relativity
(c) Photo electric cell (d) None of these.

37. ....Muscle is the strongest of all in the human body
(a) Upper arm (b) Thigh
(c) Jaw (d) Calf.

38. Elements with atomic numbers between 90 and 103 are called
(a) Lanthanides (b) Trans-uranic elements
(c) Actinides (d) Transition metals.

39. A body is in equilibrium because-
    (a) No force acts on it
    (b) A large number of forces act on it
    (c) Frictional forces act on it
    (d) Forces acting on it cancel each other.

40. Sun's heat reaches Earth by which of the following modes of heat transmission?
    (a) Convection (b) Radiation
    (c) Conduction (d) Heat exchange.

41. Vitamin A is necessary is our body for
    (a) Hemoglobin synthesis
    (b) Good digestion
    (c) Proper vision
    (d) The process of blood clotting

42. Insulin is secreted by
    (a) Thyroid gland (b) Islets of langerhans
    (c) Pituitary gland (d) Adrenal gland.

43. If an isolated conducting sphere is given a positive charge, its mass
    (a) Decreases
    (b) Increases
    (c) Remains the same
    (d) Is not involved in this case.

44. Wood is the main raw material for the manufacture of
    (a) Ink (b) Gun powder
    (c) Paint (d) Paper.

45. Which one of the following concepts represents a vector?
    (a) Kinetic emergy (b) Work
    (c) Angular momentum (d) Power.

46. The wavelenghts of X-rays lie
    (a) In the radioregion
    (b) In the infrared region

(c) In the visible region
(d) In extremely short ultraviolet region.

47. Safety fuse wire used in domestic electrical appliances is made of metal of low-
(a) Specific gravity (b) Melting point
(c) Resistance (d) None of these.

48. A chemical element in two or more forms having different physical properties but rising to identical chemical compounds is called
(a) Isomer (b) Isotope
(c) Allotrope (d) Isotone.

49. Water should not be used for extinguishing fire caused by petrol because
(a) Both water and petrol have equal-densities
(b) Water is heavier than petrol
(c) Water is lighter than petrol
(d) None of these.

50. An air bubble inside water behaves as a -
(a) Bifocal lens (b) Converging
(c) Cylindrical lens (d) Diverging lens

ANSWER SHEET

1.	(b)	2.	(c)	3.	(d)	4.	(d)
5.	(c)	6.	(a)	7.	(a)	8.	(b)
9.	(a)	10.	(c)	11.	(a)	12.	(b)
13.	(d)	14.	(b)	15.	(c)	16.	(c)
17.	(d)	18.	(d)	19.	(b)	20.	(b)
21.	(d)	22.	(c)	23.	(c)	24.	(c)
25.	(c)	26.	(c)	27.	(a)	28.	(d)
29.	(b)	30.	(a)	31.	(d)	32.	(b)
33.	(b)	34.	(a)	35.	(a)	36.	(c)
37.	(b)	38.	(c)	39.	(d)	40.	(b)
41.	(c)	42.	(b)	43.	(c)	44.	(d)
45.	(c)	46.	(d)	47.	(b)	48.	(c)
49.	(b)	50.	(d)				

SET V
ASSESS YOURSELF

1. Saline Soil contains-
   (a) A lot of moisture
   (b) High concentration of Salts.
   (c) Hard rocks
   (d) None of these.

2. Ottohan discovered atom bomb by the principle of-
   (a) A – uranium fission (b) Radiation
   (c) Nuclear fission (d) B radiation.

3. Bamboo is a
   (a) Shrub (b) Tree
   (c) Grass (d) Herb.

4. Black hole of the final density of all stars, was predicated by
   (a) Albert Einstein (b) Issac Newton
   (c) Copernicus. (d) Eratosthenes.

5. A child whirls a stone tied to a string keeping himself at the centre. The stone revolves in a circular motion due to-
   (a) Inertia
   (b) Contrifugal force
   (c) Horizontal force
   (d) None of these.

6. The gas first enters blood and then leaves it by the process of
   (a) Diffusion (b) Osmosis
   (c) Condensation (d) Transpiration.

7. Aryabhatta discovered
   (a) Exact value of
   (b) Perfect definition of Zero.
   (c) Rule of solution of simple inder-terminate equations.
   (d) None of these.

8. Newton's Rings are
   (a) Rings presented to Newton by a King.
   (b) Elastial bodies named after Newton.
   (c) An optical phenomenon.
   (d) Coloured rings observed round the point of contact of a convex lens.

9. The cheapest method of protecting foodgrains from the attack of fungi is
   (a) Drying the food grains in sunlight.
   (b) Burying the foodgrains in the ground.
   (c) Spraying the foodgrains with insecticides.
   (d) Fumigation.

10. Liquid is characterised by-
   (a) No fixed shape or volume.
   (b) Fixed shape and volume
   (c) High compressibility.
   (d) Fixed volumes.

11. The blood-sucking organisms are-
   (a) Lecches (b) Ringworms.
   (c) Earth worms (d) Hook worms.

12. In which of the following does Nuclear fission take place?
   (a) Cell division (b) Radiation
   (c) Sun (d) None of the above.

13. Laperoscopy is associated with
   (a) Study of laser beams
   (b) Science of tissues
   (c) Radars
   (d) A gynaecological operation.

14. Camphor, when exposed to air-
    (a) Forms a liquid
    (b) Sublimates
    (c) Evaporates
    (d) Turns black and forms anaxide.

15. When we go upward, the atmospheric temperature?
    (a) First increases and then decreases
    (b) Decreases
    (c) Increases
    (d) Remains the same, whatever height we may achieve.

16. The principal ore of lead is
    (a) Malachile. (b) Galena
    (c) Bauxite (d) Haematite.

17. A heart beat requires approximately
    (a) 0.8 Second (b) 0.6 Second
    (c) 0.5 Second (d) 0.7 Second.

18. The genetic code is associated with-
    (a) RNA molecules (b) Thiamine.
    (c) DNA Molecules (d) Riboflavin.

19. Electro magnetic induction is used in
    (a) Potentiometer
    (b) Generators
    (c) Galvanometer
    (d) Thermo cople.

20. Which of the following occurs in coal gas as well as in natural gas?
    (a) Butane (b) Propane
    (c) Acetylene (d) Methane.

21. The phenomena of vision is associated with vitamin
    (a) B. (b) D.
    (c) K. (d) A.

22. Equal volumes of all gases at the same temperature and pressure contain equal number of molecules". This law is called as

(a) Boyle's Law (b) Avogadro's Law
(c) Gay Lussac's Law (d) Charles Law.

23. Reducing is a reaction in which

(a) Transfer of negative valency takes place.
(b) An atom gains electrons.
(c) There is loss of electrons
(d) Gain of Negative valency takes place.

24. Rhizome is a/an

(a) Low pressure.
(b) Its weak gravitational power
(c) Absence of air
(d) None of these.

25. Life on the moon is not possible because of

(a) Low pressure
(b) Its weak gravitational power
(c) Absence of air
(d) High pressure.

26. Jaundice affects the

(a) Stomach (b) Pancreas
(c) liver (d) Kidney.

27. The jet engine is a/an

(a) External combustion engine
(b) Internal combustion engine
(c) Gas turbine
(d) None of these.

28. The boiling point of water is

(a) 212° F
(b) 100° C
(c) 80° R
(d) All temperatures mentioned above are correct.

29. Which of the following element is always present in organic compounds
    - (a) Sulphur (b) Carbon
    - (c) Nitrogen (d) Oxygen.

30. One Joule is equal to-
    - (a) 10^4 large (b) 10^5 large
    - (c) 10^6 large (d) 10^7 large.

31. Which of the following is an example of lever of the third order?
    - (a) Common Balance (b) Lemon squeezer
    - (c) Forcops (d) All of the above.

32. Which of the following is a vector quantity?
    - (a) Volume (b) Velocity
    - (c) Weight (d) Length.

33. The Unit of thͬ electric current is
    - (a) Watt (b) Ohm.
    - (c) Ampere (d) Volt.

34. The energy possessed by a body due to its position is known as
    - (a) Pressure energy (b) Kinetic energy
    - (c) Potential energy (d) Heat energy.

35. For the stability of a floating body its metacentre must be
    - (a) Above the centre of gravity of the floating body
    - (b) Below the centre of gravity of the floating body
    - (c) At the centre of gravity of the floating body
    - (d) None of these.

36. The weight of an object would be minimum when it is places at the
    - (a) Centre of the earth (b) Equator
    - (c) Poles (d) None of the above.

37. A thermometer reads 110° on a particular day in New Delhi. The reading is an the scale
    - (a) Fehrenheite (b) Centigrade
    - (c) Absolute (d) Celsius.

38. An electric bell is placed inside a bell Jar which is connected to a vacuum pump. When the bell Jar is completely exhausted
    (a) The loudness of sound will decrease
    (b) The loudness of sound will decrease
    (c) The loudness of sound will increase
    (d) No sound would be heard.

39. Farad is a Unit of
    (a) Charge (b) Potential
    (c) Capacity (d) None of these.

40. When the temperature of a metallic conduct car is increased its resistance, will
    (a) Decrease
    (b) Increase
    (c) Neither decreases nor increase
    (d) None of these.

41. Theodolite is an instrument used by
    (a) Cartographers (b) Navigators
    (c) Surveyors. (d) Pilots

42. The protein content in wheat is approximately-
    (a) 6% (b) 7%
    (c) 8% (d) 16%.

43. Platypus" is a kind of animal found in
    (a) Africa (b) Asia
    (c) America (d) Australia.

44. The Vegetation of the Thar desert consists of
    (a) Xerophytes (b) Hydrophytes
    (c) Mesophytes (d) Thallopytes.

45. Which of the following pairs is incorrect?
    (a) Roentgen X-rays
    (b) Newton Laws of Gravitation
    (c) Farady Diffusion of gases
    (d) Pasteur Bacteriology.

46. A human cell contains
    (a) 23 Chromosomes (b) 44 Chromosomes
    (c) 46 Chromosomes (d) 48 Chromosomes.

47. The electron was first identified by-
    (a). J. J. Thomson (b) Deniel Rutherford
    (c) J. Kepler (d) James Chadwick.

48. Albert Einstein got the Nobel prize for the-
    (a) Theory of Relativity (b) Photoelectric effect.
    (c) Laws of Gravitation (d) None of these.

49. Protein-calorie malnutrition causes-
    (a) Typhoid (b) Malaria
    (c) Kwashiarkor (d) Jeopatitis.

50. Small pox vaccine was discovered by
    (a) Edward Jenner (b) Robert Katch
    (c) Robert Hooke (d) Louis Pasteur.

ANSWER SHEET

1.	(b)	2.	(a)	3.	(c)	4.	(c)
5.	(b)	6.	(a)	7.	(c)	8.	(d)
9.	(a)	10.	(d)	11.	(a)	12.	(b)
13.	(b)	14.	(b)	15.	(b)	16.	(b)
17.	(a)	18.	(a)	19.	(b)	20.	(d)
21.	(d)	22.	(b)	23.	(b)	24.	(a)
25.	(c)	26.	(c)	27.	(c)	28.	(d)
29.	(b)	30.	(d)	31.	(c)	32.	(b)
33.	(c)	34.	(c)	35.	(a)	36.	(a)
37.	(a)	38.	(d)	39.	(c)	40.	(b)
41.	(c)	42.	(c)	43.	(d)	44.	(a)
45.	(c)	46.	(c)	47.	(a)	48.	(b)
49.	(c)	50.	(a)				

SET VI
ASSESS YOURSELF

1. The main function of white blood corpucles is to
   (a) Carry food
   (b) Carry oxygen
   (c) Given red colour to the blood
   (d) Produce antibodies.

2. Goitre is associated with
   (a) Thyorid gland (b) Pituitary glands
   (c) Bile (d) Parathyroid glands.

3. Which of the following substances is an important crytalline substance in the chemistry of bones and teeth?
   (a) Calcium carbonate
   (b) Gypsum
   (c) Octacalcium phosphate
   (d) Calcium sulphate.

4. The origin of species written by
   (a) Mendel (b) Lamarck
   (c) Darwin (d) Einstein.

5. A plant with green leaves placed in red light will appear
   (a) Black (b) Green
   (c) Red (d) Violet.

6. Eelectromagnetic induction is emitted by
   (a) X-rays (b) Electrons
   (c) Ultrasonics (d) Protons.

7. Which of the following animals resembles most to human being in terms of physical and mental capabilities?
   (a) Gorilla (b) Chimpanzee
   (c) Phesus monkey (d) Langur.

8. Which of the following is negligibly present in milk?
   (a) Iron (b) Fat
   (c) Protein (d) Vitamin B-complex.

9. Destruction of ecosystem leads to increase in ultraviolet radiaticons earth because of-
   (a) Increase in CO and CO_2 content
   (b) Increase in CO_2 content
   (c) Decrease in oxygen content
   (d) Decrease in ozone content

10. To drink stream water it is advisable to
    (a) Take a rip to check the presence of any salt
    (b) Take the water in a glass and put it under sun light to observe the presence of suspended particles if any
    (c) Boil the water before drinking
    (d) Filter the water with the halp of a cloth.

11. The common refrigerant used in domestic refrigerator is
    (a) Neon (b) Frion or ammonia
    (c) Alcohol (d) None of the above.

12. Which of the following animals a short tailed rodent used in scientific experimentation?
    (a) Guinea pigs (b) Rabbits
    (c) Sheep (d) Goats.

13. Who discovered the circulation of blood in the human body?
    (a) William Harvey (b) Robert Hooke
    (c) Joseph Lister (d) Andreos Gesalius.

14. Which of the following reactions is used for increasing the number of carbon atoms in a chain?
    (a) Cannizaro's reaction (b) Wurtz reaction
    (c) Friedel- Gaft's reaction (d) Substitution reaction.

15. Oxygen was discovered by-
    (a) William Rammsay (b) Rutherford
    (c) J. Priestly (d) Neil Bohr.

16. Cylindrical glasses are advised to a patient suffering from
    (a) Myopia (b) Hypermentropia
    (c) Astigmatism (d) Night blindness.

17. Feeding of colostrum important for young calves because
    (a) It is tasty
    (b) It contains antibodies
    (c) It contains iron
    (d) None of these above.

18. Why is the light of an ordinary lamp holter than the tube light?
    (a) It is due to in candesence
    (b) It is due to ionisation of gases
    (c) It has more surface of distribution of energy
    (d) None of these.

19. Which of the following is electronic?
    (a) O_2 (b) N_2
    (c) NH_3 (d) Cl_2.

20. Which of the following statement is incorrect about benzene?
    (a) All atoms lie in the same plane
    (b) Carbon and hydrogen atoms are in different planes.
    (c) All C-C bond lengths are same
    (d) All C-H bond lenghts are same.

21. Which of the following is a mineral coal?
    (a) Bauxite (b) Anthracite
    (c) Ilmenite (d) Graphite.

22. The Giant Panda which is facing extinction belongs to
    (a) Japan (b) Kenya
    (c) Indonesia (d) China

23. The most effective engine in car using
    (a) Petrol (b) Electricity
    (c) Steam (d) Diesel.

24. Bronze is an alloy of
    (a) Zinc, tin, and lead (b) Copper, lead and zinc
    (c) Copper and tin (d) None of these.

25. Night blindness is caused due to the deficiency of vitamin-
    (a) A (b) B
    (c) C (d) D.

26. Rickets is caused due to the deficiency of vitamin
    (a) A (b) B
    (c) C (d) D.

27. Coral reefs are made of-
    (a) Calcium carbonate (b) Sand of quartz
    (c) mudstone (d) None of these.

28. Cholera is caused by
    (a) Bacteria (b) Virus
    (c) Protozoa (d) Food poisoning.

29. Which of the following is an elements?
    (a) Carbon deoxide (b) Sulphur
    (c) Alcohol (d) Water.

30. Rice bran is rich in vitamin
    (a) B_1 (b) B_2
    (c) B_6 (d) B_{12}.

31. The deficiency of vitamin B_{12} causes-
    (a) Anaemia (b) Diarrhoea
    (c) Beri-beri (d) None of the above.

32. Mushroom is/an
    (a) Algae (b) Virus
    (c) Kind of fruit (d) Fungi

33. Which of the following may be caused by farmyard animals?
    (a) Glandular fever (b) Chicken pox
    (c) Poliomyelitis (d) Tuberculosis.

34. Milk is a rich source of
    (a) Riboflavin (b) Pantothenic acid
    (c) Pyridoxine (d) Thiamine.

35. Isotopes are the atoms of the elements whose nuclei contain
    (a) Same atomic number and the same atomic weight
    (b) Same number of protons but different atomic weight
    (c) Different atomic number and different number of neutrons
    (d) Different number of neutrons.

36. In the process of photosynthesis-
    (a) Oxygen is absorbed and carbon dioxide is liberated
    (b) Oxygen is liberated and nitrogen is absorbed
    (c) Nitrogen is absorbed and carbon dioxide is liberated
    (d) Carbon dioxide is absorbed and oxygen is libereted.

37. A transformer is used to
    (a) Transform AC into DC
    (b) Provide suitable DC voltage
    (c) Provide suitable AC voltage
    (d) Change the frequency of AC.

38. Amino acids are required for the metabolism of
    (a) Fats (b) Proteins
    (c) Minerals (d) Carbohydrates.

39. Leprosy is a contagious disease associated with
    (a) Skin
    (b) Skin & nerves

(c) Skin, nerves and blood vessels
(d) All organs.

40. Blue litmus paper turns.........in acids
(a) Yellow (b) Black
(c) Red (d) Violet.

41. Which of the following is used in photography?
(a) Silver bromide (b) Silver iodide
(c) Silver nitrate (d) None of the above.

42. When a ray of light is emitted through a thick glass, then its speed
(a) Increases (b) Decreases
(c) Remains same (d) None of the above.

43. Radioactivity is due to the disintegration of naturally ocurring radio isotopes. The most common type of disintegration in valves-
(a) Beta-particle emission (b) Alpha-particle emission
(c) Hamma-ray emission (d) All of the above.

44. The melting piont of which of the following elements is more than 3000°C?
(a) Cobalt (b) Beryllium
(c) Radium (d) Carbon.

45. The heads of safety matches do not contain
(a) Phosphorous (b) Sulphur
(c) Antimony trisulphide (d) Potassium chlorate.

46. Colour blindness is caused by
(a) Heredity transmission (b) High myopia
(c) Fault in the retina (d) None of the above.

47. Rentition in a human being should be complete by the time when he is.........years old.
(a) 12 (b) 18
(c) 21 (d) 24.

48. Which of the following is the most prolific breeder?

(a) Goat (b) Cat

(c) Pig (d) Bitch.

49. The main source of flouride for the human beings is-

(a) Iodized salt (b) Drinking water

(c) Vegetables &fruits (d) Alcohol.

50. Vector control is more effective in controlling

(a) Trachoma (b) Kala- zar

(c) Malaria (d) Cholera.

ANSWER SHEET

1.	(d)	2.	(a)	3.	(c)	4.	(c)
5.	(a)	6.	(b)	7.	(b)	8.	(a)
9.	(d)	10.	(c)	11.	(b)	12.	(a)
13.	(a)	14.	(b)	15.	(c)	16.	(c)
17.	(b)	18.	(a)	19.	(c)	20.	(b)
21.	(b)	22.	(d)	23.	(b)	24.	(a)
25.	(a)	26.	(d)	27.	(a)	28.	(b)
29.	(b)	30.	(d)	31.	(a)	32.	(d)
33.	(d)	34.	(d)	35.	(b)	36.	(d)
37.	(c)	38.	(b)	39.	(b)	40.	(c)
41.	(a)	42.	(b)	43.	(a)	44.	(d)
45.	(a)	46.	(c)	47.	(d)	48.	(c)
49.	(b)	50.	(c)				

SET VII
ASSESS YOURSELF

1. Fathometer is used to measure
   (a) Density of sea water
   (b) Depth of sea
   (c) Height of a mountain
   (d) None of these.

2. Unit of energy in MKS system is
   (a) Volt (b) Erg
   (c) Ohm (d) Joule.

3. Cream in milk is separated on account of:
   (a) Frictional force (b) Centripetal force
   (c) Centrifugal force (d) Gravitational force

4. In cold countries, in cold season, water pipes burst due to
   (a) Increase in volume of water
   (b) Decrease in volume of water
   (c) Increase in pressure
   (d) Great force exerted by pressure.

5. When a tuning fork of frequency 288 is sounded with another tuning fork, four beats are heard. When wax is added on the first then there is increase in the number of beats. What is the frequency of second tuning fork?
   (a) 142 (b) 292
   (c) 284 (d) 146

6. The amount of heat required to raise temperature of ....substance through 1°K is called specific heat of the substance.

(a) 1 gm (b) 1 Kg
(c) 2gm (d) 2 kg

7. Isoclinic line means

(a) Equal line (b) Equal clinic
(c) Same non-clinic (d) Same clinic.

8. Find least count of Vernier Callipers whose vernier scale is 10 and mean scale 0.9 mm.

(a) 0.09 (b) 0.9
(c) 1.1 (d) 0.1.

9. Celsius temperature corresponding to 98.4° F is

(a) 37.04°C (b) 3.72°C
(c) 17.46°C (d) 37.6°C.

10. The moment of a force is equal to

(a) Force x time (b) Mass x force
(c) Mass x time (d) Mass x acceleration.

11. Elasticity of which of the following is the largest?

(a) Rubber (b) Plastic
(c) Glass (d) Wood.

12. Which planet is the nearest to the Sun?

(a) Mercury (b) Mars
(c) Earth (d) Venus.

13. At what point the reading of Centigrade and Fahrenheit scale is equal?

(a) 40°C (b) -40°C
(c) -10°C (d) 0°C.

14. When solids change directly into gas, the process involved is called

(a) Osmosis (b) Fission
(c) Sublimation (d) Diffusion.

15. T.B. is the disease of
    (a) Heart (b) Lungs
    (c) Blood (d) None of these.

16. Enzymes are formed of
    (a) Amino acids (b) Fats
    (c) Carbohydrates (d) Proteins.

17. Yeast is used in making bread because it produces
    (a) Carbon (b) Sugar
    (c) Bacteria (d) Oxygen.

18. Which of the following gases is used in extinguishing fire?
    (a) Carbon monoxide (b) Carbondioxide
    (c) Sulphur dioxide (d) Nitrogen.

19. Food is mainly digested in
    (a) Mouth (b) Liver
    (c) Small intestine (d) Large intestine.

20. Universal donor is a person having the blood group
    (a) A (b) B
    (c) O (d) AB.

21. Man cannot survive without taking minimum amount of
    (a) Proteins (b) Carbohydrates
    (c) Fats (d) Minerals.

22. The largest gland in the human body is
    (a) Heart (b) Kidney
    (c) Liver (d) Pancreas.

23. Which of the following is a chemical change?
    (a) Melting of iron (b) Bending of iron
    (c) Rusting of iron (d) Tempering of iron.

24. Mixture of water and alcohol can be separated by
    (a) Filtration (b) Distillation
    (c) Sublimation (d) Chromatography.

25. Light energy is converted into chemical energy by
    (a) Electrolysis
    (b) Respiration
    (c) Transpiration
    (d) Photosynthesis.

26. Element present in the largest amount in human body is-
    (a) Hydrogen (b) Oxygen
    (c) Carbon (d) Nitrogen.

27. Green tie in red light appears to be
    (a) Black (b) Orange
    (c) Red (d) Yellow.

28. Both the pans of a balance are kept in an equilibrium by putting equal weights. When the balance is kept in an upward lift, the equilibrium
    (a) Will be disturbed
    (b) Will remain undisturbed
    (c) Disturbed, if the lift is accelerated
    (d) Will remain undisturbed if lift is moving with a constant velocity.

29. It takes ten minutes to increase the temperature of body from 70°C to 80°C. Time required to increase the temperature of the body from 60°C to 70°C would be.
    (a) 10 minutes
    (b) Less than 10 minutes
    (c) More than 10 minutes
    (d) Data is inadequate so can't be determind.

30. A number of images are formed when a burning candle is put before a plain mirror. The brightest image would be
    (a) First one (b) Second one
    (c) Third one (d) Last one.

31. The instrument used to measure current is
    (a) Galvanometer (b) Voltmeter
    (c) Ammeter (d) Thermometer.

32. The sky appears blue because of
    (a) Refraction of sun-rays
    (b) Reflection from the earth
    (c) Presence of sodium vapours
    (d) Presence of dust and water particles, which scatter blue colour in the higher regions.

33. On the surface of the moon, the weight of a person will
    (a) Increase (b) Decrease
    (c) Remain the same (d) Increase with time.

34. Earth is a huge magnet. This statement is
    (a) True (b) False
    (c) Partially true (d) None of these.

35. The unit of air pressure is
    (a) Volt (b) Ampere
    (c) Ohm (d) Bar.

36. Temperature of normal human body is
    (a) 98.4°C (b) 547°C
    (c) 31°C (d) None of these.

37. Newton's third law is
    (a) Action has reaction
    (b) Action is equal to reaction but in the same direction
    (c) Reaction is the effect of action
    (d) Action and reaction are equal but in opposite direction.

38. The chemical name of common salts
    (a) Sodiium chloride (b) Potassium chloride
    (c) Salt (d) Gypsum.

39. If you keep one kg cotton and one kg iron in a balance
    (a) Cotton will weigh more
    (b) Iron will weighs more
    (c) Cotton will weighs less
    (d) Both will weigh equal.

40. Water boils at
    (a) 200°F (b) 104°F
    (c) 100°C (d) 99.90 k.

41. Swimming in sea water is much easier than in river because
    (a) Density of sea water is lesser
    (b) Density of sea water is more
    (c) Density of both is same
    (d) None of these.

42. Sound is produced due to
    (a) Electrical impulses
    (b) Magnetic impulses
    (c) Electro magnetic waves
    (d) Mechanical vibration.

43. The instrument used to reduce voltage is known as-
    (a) Voltmeter (b) Transformer
    (c) Chromometer (d) Oscilloscope.

44. A cut diamond sparkles because of
    (a) Its high refractive index and high value of critical angle.
    (b) Its high refractive index and small value of critical angle
    (c) Its hardness
    (d) Its very low refractive index.

45. A motor car is moving with uniform velocity on a rough harizontal road. According to Newton's law of motion
    (a) There is no force acting on the car.
    (b) The kinetic energy of the car is increasing
    (c) Forces are acting on the car
    (d) The car is being accelerared.

46. A person standing in front of a mirror finds that his image is larger than himself. This implies that the mirror is
    (a) Convex (b) Concave
    (c) Plane (d) Plano-convex.

47. Which one of the following is an ore of iron?
    (a) Magnesite (b) Haematite
    (c) Bauxite (d) Gypsum.

48. The raw material used for the manufacture of glass and cement is
    (a) Gypsum (b) Clay
    (c) washing soda (d) limestone

49. Iron deficiency in diet causes
    (a) Anaemia (b) Demantia
    (c) Arthritis (d) Diabetes.

50. Which one of the following causes the disease AIDS?
    (a) Virus (b) Protozoa
    (c) Bacteria (d) Fungi.

ANSWER SHEET

1.	(b)	2.	(d)	3.	(c)	4.	(a)
5.	(b)	6.	(b)	7.	(d)	8.	(b)
9.	(d)	10.	(b)	11.	(d)	12.	(a)
13.	(b)	14.	(c)	15.	(b)	16.	(d)
17.	(a)	18.	(b)	19.	(c)	20.	(c)
21.	(a)	22.	(c)	23.	(c)	24.	(b)
25.	(d)	26.	(c)	27.	(a)	28.	(d)
29.	(c)	30.	(a)	31.	(c)	32.	(d)
33.	(b)	34.	(a)	35.	(d)	36.	(c)
37.	(d)	38.	(a)	39.	(d)	40.	(c)
41.	(b)	42.	(d)	43.	(d)	44.	(b)
45.	(c)	46.	(b)	47.	(b)	48.	(d)
49.	(a)	50.	(a)				

SET VIII
ASSESS YOURSELF

1. The principle of buoyancy was given by
   (a) Loues Pasteur (b) Newton
   (c) Galileo (d) Archimedes.

2. AC current can not be used in
   (a) Lighting (b) Refrigeration
   (c) Heating (d) Electro plating.

3. The temperature of the body in Kelvin scale is
   (a) 98.5 (b) 300.0
   (c) 310.0 (d) 371.0.

4. Line "spectrum" is obtained by
   (a) Burning coal (b) Sun
   (c) Filament of bulb (d) Mercury lamp.

5. Static electricity can be obtained by-
   (a) Induction only (b) Friction only
   (c) Both (a) and (b) (d) None of these.

6. Mariner's compass can not be used in
   (a) Aeroplanes (b) Satellite
   (c) Galvanometer (d) Thermopile.

7. Magnetie field is mesured by
   (a) Fluxometer (b) Pyrometer
   (c) Galvanometer (d) Thermopile.

8. That white light is made up seven colours was discovered by-
   (a) Newton (b) C.V. Raman
   (c) Galileo (d) Einstein.

9. Which of the following material has the highest refractive index
   (a) Diamond (b) Glass
   (c) Air (d) Water.

10. A rising barometer indicates
    (a) Settled fair weather
    (b) Approach of storm
    (c) Approach of fair weather
    (d) Approach of rain.

11. Water has the maximum density at
    (a) 0°C (b) 1°C
    (c) 4°C (d) 8°C.

12. Sun is continuously releasing energy due to
    (a) Nuclear Fusion (b) Radiation
    (c) Nuclear Fission (d) None of the above.

13. The surface tension is the highest for
    (a) Ether (b) Water
    (c) Alcohol (d) Mercury.

14. Fathometer is an instrument used for measuring
    (a) The density of Liquids (b) Blood pressure
    (c) Depth of the sea (d) The height of mountains.

15. Energy stored in the dry cell is
    (a) Heat (b) Chemical
    (c) mechanical (d) electrical.

16. Light year means the
    (a) Distance light travels in a year
    (b) Intensity of light
    (c) Bright light
    (d) Time light takes to travel in a year.

17. An article weighing 38.4 Kg on Earth will weigh on the Moon roughly

(a) 6.4 Kg (b) 7.5 Kg
(c) 9.6 Kg (d) 30.0 Kg.

18. An instrument for measuring the emission of radiant energy is known as

(a) Radiometer
(b) Odometer
(c) Radiator
(d) Stereoscope.

19. The instrument which projects images of opaque articles on a screen, is known as

(a) Microtome (b) Barometer
(c) Projector (d) Epidiascope.

20. J.L. Baird is credited with the invention of

(a) X-ray (b) Transistor
(c) Television (d) Telephone.

21. The direct vision spectroscope is based on the principle of

(a) Diffraction
(b) Interference
(c) Dispersion without deviation
(d) Deviation without dispersion.

22. The line joining the two poles of a magnet inside its body is called

(a) Magnetic meridian
(b) Magnetic length
(c) Magnetic axis
(d) None of the above.

23. The rod which acts as positive in a dry cell is made of

(a) Brass (b) Tin
(c) Carbon (d) Zinc.

24. Richter scale is used to measure

(a) Temperature of bodies

(b) Intensity of wind
(c) Magnitude of earthquake
(d) Intensity of earthquakes.

25. Sodium is normally preserved in
(a) Ether (b) Kerosene oil
(c) Alcohol (d) Water.

26. Which is the most abundant metal in earth's crust?
(a) Iron (b) Sodiun
(c) Silicon (d) Aluminium.

27. Electron was discovered by
(a) Planck (b) Ruther Ford
(c) Neil Bohr. (d) J. J. Thomson.

28. Which of the following is the best insulator?
(a) Ebonite (b) Paper
(c) Cotton (d) Wood.

29. Permanent hardness of water is due to the presence of
(a) Magnesium carbonate
(b) Calcium chloride
(c) Zinc chloride
(d) Calcium bicarbonate.

30. Glass is made by fusing together
(a) Soda, sand and clay
(b) Lime and clay
(c) Lime and sand
(d) Lime soda and sand.

31. Which of the following is not an alloy of tin?
(a) Brass (b) Gun meatal
(c) Solder (d) Bronze.

32. Which of the following varieties of carbon is used for making electrodes?
(a) Coke (b) Animal charcoal
(c) Gas carbon (d) Lamp black.

33. Produces gas is a mixture of
 (a) Carbon monoxide and nitrogen
 (b) Carbon dioxide and nitrogen
 (c) Carbon dioxide and hydrogen
 (d) Carbon monoxide an hydrogen.

34. Calcium carbide on treatment with water gives
 (a) Ethylene (b) Ethane
 (c) Acetylene (d) Methane.

35. An example of calloidal solution is
 (a) Brine (b) Blood
 (c) Carbonated water (d) Hard water.

36. Bleaching powder is a/an
 (a) Basic salt (b) Normal salt
 (c) Acidic salt (d) Mixed salt.

37. Plaster of Paris is
 (a) $Ca\ SO_4.\ 2H_2O$
 (b) $CaSO_4\ 5H_2O$
 (c) $Ca\ SO_4.\ H_2O$
 (d) $2\ CaSO_4.\ H_2O$.

38. Stomach in vertebrates is the main site for the digestion of
 (a) Nucleic acid (b) Fats
 (c) Proteins (d) Carbohydrates.

39. Of the following which is a plant?
 (a) Sundew (b) Sand dallar
 (c) Sea lily (d) Sea cucumber.

40. The hereditary material in both plants and animals is
 (a) DNA (b) Gene
 (c) Ribosome (d) Chromosome.

41. Nitrogen fixation is carried out by
 (a) Blue-green algae (b) Red algae
 (c) Brown algae (d) Green algae.

42. Who was the founder of bacteriology
    (a) Charles Darwin (b) De Vries
    (c) Hooke (d) Louis Pasteur.

43. Which of the following molecules is excited during photosynthesis?
    (a) Chlorophyll (b) Water
    (c) Carbon dioxide (d) Oxygen.

44. Strips on the back of the animals are indication of
    (a) Cave dwelling
    (b) Mating habits
    (c) Food habits
    (d) Matching of skin calour with surroundings.

45. Vitamin responsible for anti-sterile activity is
    (a) Vitamin A (b) Vitamin B
    (c) Vitamin E (d) Vitamin K.

46. Which of the following is the fastest growing tree?
    (a) Coconut (b) Banyan
    (c) Teak (d) Eucalyptus.

47. Which of the following does not provide minerals to the body?
    (a) Milk (b) Sugar
    (c) Fruits (d) All of these.

48. Which of the following is an improtant part of endocrine system
    (a) Spleen (b) Pituitary
    (c) Liver (d) Brain.

49. The muscles of the heart are
    (a) Involuntary (b) Valuntary
    (c) Stripped (d) Unstripped.

50. Red blood corpuscles are formed in the
    (a) Kidneys (b) Bone marrow
    (c) Heart (d) Lungs.

ANSWER SHEET

1.	(d)	2.	(d)	3.	(c)	4.	(d)
5.	(b)	6.	(b)	7.	(a)	8.	(a)
9.	(a)	10.	(c)	11.	(c)	12.	(a)
13.	(b)	14.	(c)	15.	(b)	16.	(a)
17.	(a)	18.	(a)	19.	(d)	20.	(c)
21.	(c)	22.	(c)	23.	(a)	24.	(c)
25.	(b)	26.	(c)	27.	(b)	28.	(b)
29.	(b)	30.	(d)	31.	(a)	32.	(c)
33.	(a)	34.	(c)	35.	(b)	36.	(d)
37.	(d)	38.	(c)	39.	(a)	40.	(a)
41.	(a)	42.	(d)	43.	(a)	44.	(d)
45.	(c)	46.	(d)	47.	(b)	48.	(b)
49.	(a)	50.	(b)				

SET IX
ASSESS YOURSELF

1. Which of the following combinations is/are correct?
   I. Astigmatism — Thyroid
   II. Myxodeama — Eye
   III. Acromogaly — Pitutary
   (a) I and II (b) I only
   (c) III only (d) II and III.
2. Blood in the human body is correct to the heart through
   (a) Nervous (b) Veins
   (c) Arteries (d) Muscle fibres.
3. An alloy of mercury with another metal (s) is called
   (a) Salt (b) Solution
   (c) Amalgam (d) Compound.
4. Safety wire, used in electrical circuit, is made of a material having
   (a) Low melting point
   (b) High resistance
   (c) High melting point
   (d) Low specific heat.
5. Which of the following describes the path of Halley's Comet?
   (a) Circle (b) Parabola
   (c) Hyperbola (d) Ellipse.

6. The temperature on the surface of Sun is approximately
   (a) 3000°C (b) 4000° C
   (c) 5000° C (d) 6000° C.

7. During the process of photosynthesis the oxyzen from sunlight is released by
   (a) Chlorophyll (b) Water
   (c) Carbohydrates (d) Carbon dioxide.

8. Pitutary gland is an endocrine gland attached to the
   (a) Spleen (b) Liver
   (c) Neck region (d) Base of the brain.

9. When a monocromatic light is passed through a prism, it undergoes
   (a) Dispersion (b) Diffraction
   (c) Polarisation (d) Refraction.

10. Diode is used for
    (a) Oscillation (b) Amplification
    (c) Rectification (d) Modulation.

11. A finally-cut diamond shines because of
    (a) Refraction (b) Reflection
    (c) Flourescence (d) Total internal reflection.

12. Which of the following pairs is correct?
    (a) DOT- Insectiside
    (b) Heavy Water—Moderater
    (c) Methy Alcohal-Alkohlic beverage
    (d) Liquid Ammonia—Refrigerant.

13. The tree that sends down roots from its branches to the soil is known as
    (a) Palm (b) Oak
    (c) Banyan (d) Pine.

14. Yeast is a
    (a) Fungus (b) Fern
    (c) Bacteria (d) Algae.

15. The root mean square voltage of AC main supplied in India is
    (a) 220 V (b) 110 V
    (c) 2202V (d) 220V2.

16. Which of the following diseases has been most prominent in the news in recent days?
    (a) Leprosy (b) AIDS
    (c) Cancer (d) Thalassaemia.

17. A body to move in a circular motion requires.
    (a) Inertial force (b) Centripetal force
    (c) Gravitational force (d) Centrifugal force.

18. We are able to see things because
    (a) The rays of sun fall on the objects
    (b) The light emitted by the objects reaches our eyes
    (c) The light scattered by the objects reaches our eyes
    (d) The light emitted by our eyes falls on the objects.

19. The chemical names of washing soda and caustic soda are respectively
    (a) Sodium carbonate and sodium hydroxide
    (b) Sodium bicarbonate and pottassium hydroxide
    (c) Sodium carbonate and pottassium hydroxide
    (d) Sodium bicarbonate and sodium hydroxide.

20. The nucleus of deuterium contains
    (a) One proton and one neutron
    (b) One neutron and one electron
    (c) One proton and one electron
    (d) two protons.

21. Which of the following mixtures is used for welding iron?
    (a) Oxygen and Acetylene
    (b) Nitrogen and Hydrogen
    (c) Nitrogen and Chlorine
    (d) Oxyzen and Nitrogen.

22. Distant objects can be seen with the help of
    (a) Telescope (b) Oscilloscope
    (c) Anemometer (d) Spectroscope.

23. Whiteness of milk is due to the presence of
    (a) Lectose (b) Cacotin
    (c) Cesin (d) Albumin.

24. In photosynthesis light energy is transformed into
    (a) Heat energy (b) Chemical energy
    (c) Mechanical energy (d) Atomic energy.

25. Hydrogan bomb works on the principle of
    (a) Nuclear fission (b) Nuclear fusion
    (c) Radiation (d) Nuclear reaction.

26. The velocity of a particle is 20ft/sec and accileration is 10ft/ sec^2. What is the distance traversed by the particle after 20 second
    (a) 240 ft (b) 2400 ft
    (c) 4000 ft (d) 24000 ft.

27. If the length of a simple pendulum is increased then the time period will
    (a) Decrese (b) Increase
    (c) Remain the same (d) None of these.

28. A man weighing 50kg is moving round a radius of 6 metres. What is the work done
    (a) 300 ergs (b) 3000 ergs
    (c) Zero (d) 250 joules.

29. Hardness of water is due to the presence of
    (a) Bicarbonates of calcium and megnisium
    (b) Sulphate of calcium and megnisium
    (c) Nitrates of calium and megnisium
    (d) Chlorides of calium and megnisium.

30. The nucleus of the atom contains
    (a) Neutrons (b) Proton
    (c) Proton and neutrons (d) Electrons and protons.

31. The cathode rays are identical with

(a) A-particle (b) B-particle
(c) Γ-rays (d) A hydrogen.

32. If two sounds heard at the same time and place have the same wavelength, they must have the same

(a) Amplitude (b) Frequency
(c) Quality (d) Loudness

33. Mercury thermometer can not be used at 360°C temperature because

(a) Boiling point of mercury is 352°C
(b) Mercury breaks the tube
(c) Mercury is not clearly visible at that tempreture
(d) None of these.

34. If two long prallel wires have currents flowing in the opposite direction then

(a) There will be a force of attraction between the wires
(b) There will be a force of repuls on between the wires
(c) Both (a) and (b)
(d) None of these .

35. Common element in all the organic compounds other than H_2 is

(a) Nitrogen (b) Sulfur
(c) Phosphorus (d) Carbon.

36. Light waves are

(a) Longitudinal waves
(b) Transverse waves
(c) Resonant
(d) both (a) and (b).

37. The focal length of a concave mirrors depends upon

(a) The object distance from the mirror
(b) The image distance of the mirror
(c) Both (a) and (b)
(d) None of these.

38. The power of a lens is inversely proportional to the
    (a) Image (b) Focal length
    (c) Centre of curvature (d) None of these.

39. The branch of Botany which deals with the study of stages is
    (a) Physiology (b) Ecology
    (c) Morphogenesis (d) Genetics.

40. Stevenson screen is a
    (a) Device for measuring atmospheric pressure
    (b) Device for measuring rainfall
    (c) Screen for projecting films
    (d) Standard shelter for thermometers.

41. All living objects consume oxygen for their respiration. Oxygen is replenished by plant during
    (a) Evening (b) Night
    (c) Morning (d) Day time

42. Potato is a
    (a) Modified steam (b) Modified seed
    (c) Modified root (d) Modified fruit.

43. The micro-organism which can not multiply outside living cells are
    (a) Virus (b) Fungi
    (c) Yeast (d) Bacteria.

44. Night blindness is caused by the deficiency of
    (a) Vitamin A (b) Vitamin B
    (c) Vitamin C (d) Vitamin D.

45. What is the value of the absolute zero?
    (a) 0°C (b) -273°C
    (c) -273°F (d) -273°F.

46. What phenomenon is responsible for the formation of a rainfall?
    (a) Polarisation (b) Refraction
    (c) Dispersion (d) Reflection.

47. A rock is taken from the surface of the earth to the moon what changes will be observed ?
    (a) The mass and weight of the rock will remain the same
    (b) The mass of the rock will change
    (c) The mass and weight of the rock will change
    (d) The weight of the rock will change.

48. The unit of force in the standard international system is
    (a) Newton (b) Pound-weight
    (c) Poundal (d) Dyne.

49. The most abundant element in earth is
    (a) Iron (b) Silicon
    (c) Oxygen (d) Ntrogen.

50. The study of growth of body is called
    (a) Ecology (b) Genetics
    (c) Physiology (d) Biology.

ANSWER SHEET

1.	(c)	2.	(c)	3.	(c)	4.	(a)
5.	(c)	6.	(d)	7.	(b)	8.	(d)
9.	(a)	10.	(c)	11.	(d)	12.	(c)
13.	(c)	14.	(a)	15.	(a)	16.	(b)
17.	(b)	18.	(c)	19.	(a)	20.	(a)
21.	(a)	22.	(a)	23.	(a)	24.	(b)
25.	(a)	26.	(b)	27.	(b)	28.	(c)
29.	(a)	30.	(c)	31.	(c)	32.	(b)
33.	(a)	34.	(b)	35.	(d)	36.	(b)
37.	(c)	38.	(b)	39.	(c)	40.	(d)
41.	(d)	42.	(a)	43.	(a)	44.	(a)
45.	(b)	46.	(c)	47.	(d)	48.	(a)
49.	(c)	50.	(c)				

SET X
ASSESS YOURSELF

1. Which one of the following diseases is caused by virus?
   (a) Sleeping sickness (b) Diptheria
   (c) Small pox (d) Typhoid.

2. A body fully immersed in a fluid experiences an upward thrust which depends on
   (a) The density of the body
   (b) The volume of the body
   (c) The weight of the body
   (d) None of these.

3. When water is filled in a bottle and is allowed to freeze the bottle breaks because
   (a) Water expands on freezing.
   (b) Temperature outside the bottle is less than inside the bottle
   (c) Bottle contracts at freezing point
   (d) None of these.

4. If the distance between two similar charge separated by a unit distance, is increased five -times, than the force of repulsion will be
   (a) 1/5 (b) 2/5
   (c) 1/25 (d) 5 times.

5. Potentiometer is an ideal instrument for measuring potential difference because-
   (a) It does not disturb the potential difference it measures.
   (b) End corrections are eliminated
   (c) It has a long wire
   (d) All of the above.

6. The age of a tree can be estimated by counting-
   (a) The number of barks
   (b) The number of branches
   (c) The number of growth rings
   (d) The number of leaves.

7. The structural changes in a mamalian uterus during reproductive cycle are primarily caused by-
   (a) Thyroxine
   (b) Testestosterone
   (c) Progesterone and estrogen
   (d) Epinephrine.

8. Water boils at a lower temperature than 100°C on a hill station because
   (a) Pressure is low at high altitude
   (b) There is cloud formation at high altitudes
   (c) Water vapours are less at high altitude
   (d) Temperature is low at high altitude.

9. Frost is most likely to occur on a
   (a) Cloudy, cold night
   (b) Cold , clear night
   (c) Warm, clear night
   (d) Cloudy, warm night.

10. Of the various forms of carbon, the one which is used for absorption purpose is
    (a) Coke (b) Wood charcoal
    (c) Graphite (d) Diamond

11. The altitudes of heavenly bodies appear to be greater than they actually are. This is due to
    (a) Diffraction
    (b) Atmospheric refraction
    (c) Atmospheric reflection
    (d) Dispersion

12. Food gets cooked quickly in a pressure cooker because-
    (a) The boiling point of a liquid is not proportional to the pressure on its surface
    (b) Of high pressure inside, thereby raising the boiling point of water
    (c) Its lid is lightly-fitted
    (d) None of these.

13. A person standing before a furnace relieves most of the heat by
    (a) Convection
    (b) Conduction
    (c) Radiation
    (d) Conduction and convectiocn.

14. Water in an earthen pot cools below the room temperature due to
    (a) Absence of radiation
    (b) Evaporation of water from the surface of the pot
    (c) Insulation
    (d) Absence of convection.

15. Water has maximum density at
    (a) -4°C (b) 4°C
    (c) 0°C (d) 100°C.

16. An object executes simple harmonic motion with amplitude A its acceleration will be maximum when the displacement is
    (a) A/4 (b) O
    (c) A/2 (d) A.

17. When load is lifted from the ground to the certain height some work is done energy is used up. This energy is converted into
   (a) Heat and dissipated
   (b) Internal energy of the body
   (c) Potential energy of the body
   (d) Heat as well as potential energy.

18. Work done in taking a mass of 10 kg up a frictionless inclined plane making an angle of 30°. with the horizontal through a distance of 10 cm is about
   (a) 5 J (b) 10 J
   (c) 20 J (d) 50 J.

19. Which one of the following species has the same number of electrons as an atom of organ?
   (a) S– (b) Na^+
   (c) Al^{+++} (d) Mg^{++}.

20. The product obtained when vegetable oil or fat is subjected to alkali treatment is
   (a) Glycrine (b) Glycerol
   (c) Glycol (d) Glayoxime.

21. The type of hybridisation of carbon atom is
   (a) Sp^2 (b) Sp
   (c) Sp^3 (d) dsp^2.

22. The O-C-O angle in CO_2 molecule is
   (a) 98° (b) less than 90°
   (c) 120° (d) 180°.

23. Phobas is a sattelite of the planet
   (a) Mass (b) Uranus
   (c) Jupiter (d) Venus.

24. What is the name of the computer which was first used for programming and playing of music ?
   (a) IILIAC (b) ENIAC
   (c) UNIVA (d) PARAM.

25. Who invented revolver ?
    (a) Reifel (b) Alfred Noble
    (c) Samuel Colt (d) Charles Patton.

26. Who is credited with the invention of basket ball ?
    (a) James Naismith
    (b) Max Planck
    (c) Sam Goldwyn
    (d) Charles Pattron.

27. Where is the world did the first motor car race take place?
    (a) USA (b) Germany
    (c) UK (d) France.

28. The Uri Hydro electric project is located in ?
    (a) J & K (b) Himachal Pradesh
    (c) Uttar Pradesh (d) Haryana.

29. The coil of an electric heater is made up of
    (a) Copper (b) Nichrome
    (c) Platinum (d) Tungsten.

30. Colour which will be best deviated by a prism is
    (a) Indigo (b) Blue
    (c) Violet (d) Red.

31. Light year is a unit of
    (a) Velocity (b) Distance
    (c) Time (d) Acceleration.

32. Sound does not travel in
    (a) Air (b) Water
    (c) Complete vaccum (d) Copper wire.

33. Beats are produced as a result of
    (a) Diffraction
    (b) Constructive interferences
    (c) Destructive interferences
    (d) Alternative constructive and destructive interferences.

34. Whose voice has greater pitch ?
    (a) Mosquito (b) Man
    (c) Woman (d) Lion.

35. The power of a lens is expressed in
    (a) Metres (b) Dioptres
    (c) Number (d) Watt.

36. Which among the following gases is filled in balloons ?
    (a) Hydrogen (b) Helium
    (c) Oxygen (d) Carbon dioxide.

37. Which of the following is used for digestion of food ?
    (a) Enzymes (b) Insuline
    (c) Harmones (d) HCl.

38. Green tie in the red light appears to be
    (a) Black (b) Orange
    (c) Red (d) Yellow.

39. Which among the following is a good conductor of electricity?
    (a) Zinc (b) Copper
    (c) Mercury (d) Silver.

40. Who discovered hydrogen gas ?
    (a) Cavendish (b) Priestley
    (c) Canbur (d) None of these.

41. Which of the following has the highest calories per grams
    (a) Proteins (b) Fats
    (c) Sugar (d) Glucose.

42. Dinosour is
    (a) An amphibian (b) A reptiles
    (c) A mammal (d) A bird

43. The chief ingredient of the mosquito repellent cream is derived from
    (a) Tulsi (b) Rice bran
    (c) Neem (d) Lemon.

44. Which of the following is not an alloy?
   (a) Brass (b) Bronze
   (c) Steel (d) Zinc.

45. Which of the following plants is biennial ?
   (a) Banana (b) Pineapple
   (c) Carrot (d) Jackfruit.

46. Sulpha drugs may be effective against diseases caused by
   (a) Parasites (b) Viruses
   (c) Bacteria (d) Worms.

47. Three basic colours in colour television are
   (a) Blue, green , yellow
   (b) Red, blue, green
   (c) Red, green, yellow
   (d) Red, blue, yellow

48. Which part of the tea plant is used for making tea ?
   (a) Root (b) Flower
   (c) Stem (d) Leaver.

49. Which of the following is not a fish?
   (a) Dog fish (b) Gold fish
   (c) Star fish (d) Jelly fish.

50. The edible part in potato is its
   (a) Bud (b) Stem
   (c) Root (d) Fruit.

ANSWER SHEET

1.	(c)	2.	(b)	3.	(a)	4.	(c)
5.	(b)	6.	(c)	7.	(c)	8.	(a)
9.	(b)	10.	(b)	11.	(b)	12.	(b)
13.	(d)	14.	(b)	15.	(b)	16.	(d)
17.	(c)	18.	(b)	19.	(a)	20.	(b)
21.	(c)	22.	(d)	23.	(a)	24.	(d)
25.	(c)	26.	(a)	27.	(c)	28.	(a)

29.	(b)	30.	(d)	31.	(b)	32.	(c)
33.	(d)	34.	(a)	35.	(b)	36.	(a)
37.	(a)	38.	(a)	39.	(d)	40.	(a)
41.	(a)	42.	(b)	43.	(c)	44.	(d)
45.	(c)	46.	(c)	47.	(b)	48.	(d)
49.	(d)	50.	(c)				

SET XI
ASSESS YOURSELF

1. Rusting of iron involves
   (a) Reduction (b) Oxygen
   (c) Decomposition (d) Displacement.
2. "Malaria" spread through the vector
   (a) Culex mosquito
   (b) Anapheles mosquito
   (c) Acedes mosquito
   (d) Tse-tse mosquito.
3. Gorowth in animals is
   (a) Indeterminate (b) Indefinite
   (c) Definite (d) Continuous.
4. The cheap source of protien of good nutritive value is found in
   (a) Rice (b) Soyabeane
   (c) Wheat (d) Oats.
5. Which one is a good preservative of food?
   (a) Sugar (b) Formaldehyde
   (c) Spirit (d) None of these.
6. Red blood corpuscles are formed in the
   (a) Heart (b) Liver
   (c) Spleen (d) Bone marrow.

7. The inner layer of the skin is called?
   (a) Dermis (b) Cuticle
   (c) Epidermis (d) None of these.

8. The dense mass of small water droplets on smoke or dust particles in the lower layer of the atmosphere is
   (a) Frost (b) Fog
   (c) Dew (d) Hail.

9. The four major components of the atmosphere are
   (a) Oxygen, hydrogen, carbon dioxide and neon
   (b) Oxygen, nitrogen, carbon dioxide and argon
   (c) Oxygen, nitrogen, carbon dioxide and hydrogen
   (d) Oxygen, nitrogen, carbon dioxide and water vapour.

10. The crop associated with retting process is
    (a) Jute (b) Cotton
    (c) Rice (d) Wheat.

11. Which of the following is not a byproduct of coal gas?
    (a) Coal tar (b) Coke
    (c) Sulphuric acid (d) Cobalt.

12. The study of fossils is known as
    (a) Palaeontology (b) Seismology
    (c) Petrology (d) Bathymetry.

13. Grafting involves the formation of a young plant
    (a) By stem cutting
    (b) With the root of stalk and shoot of scion
    (c) With the root of scion and shoot of stalk
    (d) By layering.

14. Leaves appear green because they
    (a) Scatter green light (b) Reflect green light
    (c) Absorb green light (d) None of these.

15. The hardest part in a tooth is
    (a) Dentine (b) Dental tubule
    (c) Enamel (d) Pulp.

16. Ethylene gas is used to
    (a) Ripen the fruit
    (b) Destroy the fruit
    (c) Preserve the fruit
    (d) Increase respiration in fruit.

17. Enzymes are made of substance that are nature of
    (a) Fats (b) Carbohydrates
    (c) Lipids (d) Proteins.

18. Which part of the cinnamon plants is used as a spice
    (a) Dried flowers buds (b) Bark
    (c) Rhizome (d) Stem.

19. BHC (Benzene hexachloride) is a / an
    (a) Herbicide (b) Insecticide
    (c) Fungicide (d) Weedicide.

20. If the lense in the eye becomes opaque the disease is called
    (a) Myopia (b) Gloucoma
    (c) Astigmatism (d) Cataract.

21. Which of the following disease is caused by virus?
    (a) Tetanus (b) Influenza
    (c) Diptheria (d) Syphils.

22. Which of the following is the lightest gas ?
    (a) Hydrogen (b) Carbon dioxide
    (c) Nitrogen (d) Ammonia.

23. The metal that is present in photo films is
    (a) Magnesium (b) Platinum
    (c) Silver (d) Mercury.

24. The fact which support the idea that viruses are living is that they
    (a) Can be crystallised
    (b) Are made up of common chemicals
    (c) Duplicate themselves
    (d) Pentrate cell membrane.

25. Yawning is due to
    (a) Accumulation of carbon dioxide in the body
    (b) To ease digestion
    (c) To feel relaxed
    (d) To word off lethargy.

26. Insects that make a elicking sound are
    (a) Flies (b) Beetles
    (c) Crickets (d) Silver fish.

27. Which one of the following disease is not caused by bacteria ?
    (a) Leprosy (b) Tuberculosis
    (c) Typhoid (d) Measles.

28. An enzymes that works in on acidic medium is
    (a) Ptyalin (b) Pepsin
    (c) Trypsin (d) Maltose.

29. Regeneration is a process by which
    (a) Old individuals are damaged
    (b) New individuals are produced
    (c) The individuals grows double its size
    (d) The warm out parts of the individuals are replaced

30. Water exposed to the atmospheric is slightly acidic because of
    (a) Oxygen (b) Carbon dioxide
    (c) Nitrogen (d) Suns rays

31. Potassium permanganate is
    (a) A base (b) An acid
    (c) A reducing agent (d) An oxidising agent

32. Pituitary gland is present
    (a) Below the brain
    (b) Inside the brain
    (c) Above the brain
    (d) Nowhere near the brain

33. Glucose appears in the urine in

(a) Diabetes mellious (b) Goiter

(c) Dwarfism (d) Tetany.

34. Which of the following combinations is recommended in fertilizer?

(a) Nitrogen, iodine, sodium

(b) Nitrogen, phosphorus

(c) Nitrogen, sulphur, sodium

(d) Nitrogen, sulphur, potassium.

35. The vitamin necessary for clotting of bloods is

(a) B (b) A

(c) C (d) K.

36. When water freezes its density

(a) Remains constant (b) Decreases

(c) Increases (d) Becomes zero.

37. Fishes breaths with

(a) Gills (b) Skins

(c) Lungs (d) Mouth.

38. Which of the following is the best source of proteins ?

(a) Butter (b) Apples

(c) Beans (d) Fish.

39. Which is a proper food-chain showing a producer a herbivere and a carnivore

(a) Grass - Insect- Elephant

(b) Fish-Insect-Whale

(c) Plants - Rabbit- Tiger

(d) Tiger - Rabbit - Owl.

40. Sleeping under the trees at night time is harmful but not during the day time because

(a) During the night time plants give out carbon dioxide

(b) Trees give out a harmful secretion at night time

(c) Birds that come to rest may excrete

(d) None of these.

41. A microscpic organism playing a vital vole in nitrogen cycle is

(a) Bacteria (b) Protozoa

(c) Virus (d) Red algae.

42. In our body the tissue are bathed in a fluid called

(a) Water (b) Blood

(c) Lymph (d) Plasma.

ANSWER SHEET

1.	(b)	2.	(b)	3.	(c)	4.	(b)
5.	(b)	6.	(d)	7.	(a)	8.	(b)
9.	(d)	10.	(a)	11.	(d)	12.	(a)
13.	(b)	14.	(b)	15.	(c)	16.	(a)
17.	(d)	18.	(b)	19.	(b)	20.	(b)
21.	(b)	22.	(a)	23.	(c)	24.	(c)
25.	(a)	26.	(b)	27.	(d)	28.	(b)
29.	(d)	30.	(c)	31.	(c)	32.	(a)
33.	(a)	34.	(b)	35.	(d)	36.	(b)
37.	(a)	38.	(c)	39.	(c)	40.	(a)
41.	(a)	42.	(c)				